Smart Sensors, Measurement and Instrumentation

Volume 35

Series Editor

Subhas Chandra Mukhopadhyay
School of Engineering
Macquarie University
Sydney, NSW
Australia

The Smart Sensors, Measurement and Instrumentation series (SSMI) publishes new developments and advancements in the fields of Sensors, Instrumentation and Measurement technologies. The series focuses on all aspects of design, development, implementation, operation and applications of intelligent and smart sensors, sensor network, instrumentation and measurement methodologies. The intent is to cover all the technical contents, applications, and multidisciplinary aspects of the field, embedded in the areas of Electrical and Electronic Engineering, Robotics, Control, Mechatronics, Mechanical Engineering, Computer Science, and Life Sciences, as well as the methodologies behind them. Within the scope of the series are monographs, lecture notes, selected contributions from specialized conferences and workshops, special contribution from international experts, as well as selected PhD theses.

Indexed by SCOPUS and Google Scholar.

More information about this series at http://www.springer.com/series/10617

Md Eshrat E. Alahi ·
Subhas Chandra Mukhopadhyay

Smart Nitrate Sensor

Internet of Things Enabled Real-Time Water Quality Monitoring

Springer

Md Eshrat E. Alahi
Shenzhen Institute of Advanced
Technology, Chinese Academy of Sciences
Shenzhen University Town
Shenzhen
Guangdong, China

Subhas Chandra Mukhopadhyay
School of Engineering
Macquarie University
Sydney, NSW, Australia

ISSN 2194-8402 ISSN 2194-8410 (electronic)
Smart Sensors, Measurement and Instrumentation
ISBN 978-3-030-20094-7 ISBN 978-3-030-20095-4 (eBook)
https://doi.org/10.1007/978-3-030-20095-4

This Springer imprint is published by the registered company Springer Nature Switzerland AG
The registered company address is: Gewerbestrasse 11, 6330 Cham, Switzerland

Preface

Nitrate-N is a naturally occurring ionic compound that is a part of nature's nitrogen cycle. Nitrate-N are readily lost to ground and surface water as a result of intensive agriculture, industrial wastes, and disposal of human and animal sewage. The impact of elevated nitrate-N concentrations on water quality has been identified as a critical issue of a healthy environment for the future. Presently, water quality managers follow the traditional measurement systems that involve physically collecting the sampling water from remote sites and testing it in the laboratory. These methods are expensive, require trained people to analyse the data and produce much chemical waste.

Therefore, low-cost ion-imprinted polymer (IIP)-coated impedimetric nitrate-N sensor was developed, and the detection range of nitrate-N was 1–10 (mg/L). The selective IIP material was sensitive to nitrate-N ions in an aqueous medium, and the results are validated through standard UV spectrometric methods. Micro-electromechanical system (MEMS)-based interdigital sensor and sensing system was also developed to measure nitrate-N, and the range was 0.01–0.5 (mg/L). The graphite/PDMS-based low-cost sensor was also fabricated, and the sensor was characterized to measure nitrate-N in the range of 1–70 (mg/L). Temperature compensation was added for both the sensors (MEMS and graphene), and Wi-fi connectivity was provisioned in the system to transfer the measured data in real time. An improved LoRa-based sensing system (solar-panel- and rechargeable-battery-powered) was developed and trialled in the field successfully which can measure the nitrate-N concentration in real time and transfer the data to IoT cloud server to overcome the limitations of laboratory-based sensing system.

Chapter 1 discusses the background and the introduction of the research. The importance of nitrate sensor and in-situ-based sensing system with real-time monitoring has been discussed. There are varieties of detection methods available for nitrate. Some of them are laboratory-based, and some of them can be used in the sampling locations. The advantages and shortcomings of those methods are discussed in Chap. 2. Chapter 3 discusses the principle of interdigital capacitive sensor, and the detection method of electrochemical impedance spectroscopy (EIS). Chapter 4 discusses the low-range nitrate measurement, the temperature compensation technique which is employed in the developed sensor and the electronics of the IoT-enabled smart sensing system. Chapters 5, 6 and 7 discuss the varieties of the sensor and their fabrication technique, validation methods and development of the in situ IoT-enabled sensing system. Chapter 8 discusses the conclusion of the research, limitations of the current research and prospects.

This book is originated from a Ph.D. research done at Macquarie University, NSW, Australia. The aim of this work is to develop low-cost IoT-enabled smart nitrate sensor for the detection of nitrate as contamination in natural water. The authors are highly grateful to the colleagues who had a significant contribution to this work: Dr. Anindya Nag, Dr. Nasrin Afsarimanesh, Dr. Krishanthi Jayasundera, Dr. Keith Imrie, Dr. Fahmida Wazed Tina and Van Nguyen Thi Phuoc. Special thanks to Macquarie University, Australia, and Massey University, New Zealand, for providing research facilities. We would also extend our thanks to our families for their support, motivation and encouragement throughout the work.

Shenzhen, China Dr. Md Eshrat E. Alahi
Sydney, Australia Prof. Subhas Chandra Mukhopadhyay

Contents

Chapter 1
Introduction

1.1 Introduction

Nitrogen is the most common natural element in the atmosphere, being nearly 80% of the air we breathe [1]. Nitrogen can be found in gaseous form in the air, such as Nitrogen (N_2), Nitrous oxide (N_2O), Nitric oxide (NO), Nitrogen dioxide (NO_2), and Ammonia (NH_3) [2]. Some of these gases react with rainwater and produce nitrate and ammonium ions, which can become part of the soil layer, or mix with groundwater in solution. Intake of the nitrate ion has several positive aspects for the human body, such as improved blood flow, reducing blood pressure, and cardio-vaso-protective effects. However, adverse effects can occur to the human body with an excessive intake of nitrate ions, especially through drinking water, such as gastric, cancer and Parkinson's diseases. Infants can suffer "blue baby syndrome" or methemoglobinemia [3] which reduces the oxygen content of the blood [4, 5]. It affects those infants who are less than six months old.

The Nitrate ions have been successfully utilized in some of our various activities, but there is no doubt that our fondness for them has diminished in recent years due to their excessive use. The excessive and continuous use of nitrate has caused enormous problems and raised numerous concerns [6, 7]. These problems have been identified widely all over the world, and as a result, different international and government organisations have created frameworks to control the level within the environment and in food products, with appropriate regulations in most industrialised countries [3, 8].

The need and desire to monitor nitrates is undisputable yet monitoring their presence can present a substantial challenge to the research community. Similar ions are present, such as nitrite, ammonium, phosphate, and sulfate. Therefore, improved detection methods are essential to avoid any interference that can be encountered in the environment, industries, food, and industrial activities. Many analytical methods and sensing methods have been developed to overcome the peculiarities of various media. This chapter provides a relative review of the reported literature covering all kinds of detection of nitrates and provides a brief assessment of the comparative mer-

Md E. E. Alahi and S. C. Mukhopadhyay, *Smart Nitrate Sensor*, Smart Sensors, Measurement and Instrumentation 35, https://doi.org/10.1007/978-3-030-20095-4_1

its of each detection approach. In recent years, many reviews have been undertaken to investigate the different detection methods [9–13].

In our daily life, nitrate is a vital ion, and various nitrogen species commonly occur in the environment. The excessive use of fertilisers in agriculture and general mismanagement of the use of natural resources is responsible for the perturbation of the global and local nitrogen cycle [14]. Waste materials can be one anthropogenic source of nitrates in groundwater. Disposal of animal sewage, industrial waste, and excessive use of organic fertilisers in agriculture are significant sources of nitrate pollution for water [15–17]. Nitrate in an aquatic medium stimulates excessive production of algae and phytoplankton, which leads to eutrophication. They consume more oxygen during the decomposition process that affects fish or other aquatic life. Therefore, monitoring the environmental fate of nitrates has gained increasing importance.

Recently, pollutant removal from water using membrane technology has received considerable attention. It has a high adsorption capacity with low cost and has been integrated into sensor technology when the selectivity towards the targeted ions is enhanced [18]. Appropriate characteristics and selection of the membrane material are essential in order to produce highly selective materials and systems. Many research studies have reported different types of the membrane for the detection of nitrate in water. Modified silica polymer, trihexyltetradecyl-phosphonium chloride polymeric membrane [19], and doped polypyrrole, zinc (II) complex polymeric membrane [20], have been used for membrane development. Much research is going on to improve the sensitivity of membrane materials and to improve the selectivity. In natural water, there are other environmental impacts such as temperature, vibration and the presence of other polluting ions, which will affect the precision of the measurements. All these environmental impacts might come naturally or via industrial and experimental waste. Therefore, obtaining reliable data is essential and can be achieved by controlling the measurement conditions. Precision instrumentation and measurement methods are essential to get highly reliable data. Different methods are available to detect nitrates in water, such as chromatography, flow injection analysis, electrochemical sensors, biosensors, optical fibre sensors, electromagnetic sensors. All of them have different characteristics and sensitivities.

This chapter and Chap. 2 are explained the Introduction and Literature review of the nitrate sensor, background of the research, different detection methods, their advantages and disadvantages. The discussion has been done elaborately.

In Chap. 3, capacitive interdigital sensors are designed and developed. The working principle of the interdigital sensor has been explained in this chapter. Electrochemical Impedance Spectroscopy (EIS) method is used to detect nitrate in water. The relationships of those electrodes with impedance measurement and the technique of data acquisition are also discussed in this chapter.

Chapter 4 is the first experiment in this book. A temperature compensated nitrate sensor has been developed, and experimental setup, results, and discussions are included in this chapter. The sensor was used to measure low concentration nitrate (0.1–0.5 ppm), and temperature compensation was added to improve the results. The smart IoT enabled sensing system was also explained in this chapter. The real-time data collection was shown in laboratory-based setup and validated the results by comparing with standard laboratory method. The system has the potential to use in the real sampling location. It has also reported an Internet of Things (IoT) based smart sensing system which can measure nitrate-N in real time. The developed system is a low cost (the estimated amount of the whole system is < USD 100). The availability of low-cost sensors with selectivity provides the opportunity to develop an extensive distributed network, for real-time monitoring.

In Chap. 5 showcases the irradiation of a polymer film by a laser sintering technique to generate conductive material to design and develop a nitrate sensor. The last decade has experienced unprecedented growth in the use of graphite due to its exceptional electrical and mechanical properties. As a result of its distinctive advantages over other conductive materials, graphite has become a popular choice to develop the electrodes of a sensor. Two of the significant advantages of using graphite for electrodes are its porosity and corrosion-resistant nature. The former characteristic causes an increase in the sensing area of the sensor, whereas the latter prolongs the efficiency of the sensor. The applications of graphite have been in different sectors, primarily in healthcare, environmental and industrial areas. Due to the high electrical conductivity of the graphite film generated by this technique, its use has been mostly in micro-supercapacitors to reduce the effect of polarisation at high frequencies. Less sample preparation time, smooth cuts and straight edges are some of the advantages provided by the laser sintering technique. A minimised number of steps generating porous conductive material with enhanced electrical, mechanical and thermal properties are some of the advantages of the proposed sensor. In this chapter, the fabrication process of Carbon Nano-tubes (CNT)—Polydimethylsiloxane (PDMS) and Graphite sensors.

Graphite sensors are low cost, which would be useful to measure nitrate-N concentration in water. An Internet of Things (IoT)enabled smart sensing system was also developed to measure the nitrate-N concentration in real time. IoT refers to the internet-connected network objects from everyday life, which is often capable of doing regular activity cleverly. The purpose of using the IoT is to create an environment in which the necessary information of the developed network is shared in real time. The proposed system can also measure the nitrate-N concentration in real time and transfer the data to the cloud server to allow the system to be used as part of a distributed network. Compensation for temperature was also included in the system, which will increase the accuracy of measurement. The fabrication process, nitrate-N and temperature measurement and validation of the smart sensing systems are explained in the subsequent sections.

In Chap. 6, selectivity has been introduced and biological approaches such as enzymes, whole cells and anti-body-based biosensors, have also been used for nitrate-N analysis [21–23]. Though enzymatic biosensors have a comparatively high limit of detection (LOD), they require the control of pH and the separation process of the enzyme from bacteria which is time-consuming. Immunoassay-based biosensors have reasonable sensitivity, with good selectivity, compare to the enzymatic biosensor. Despite this, there are complications with this method. The materials used for these biosensors are expensive and difficult to handle. Recently, research has focussed on the development of voltammetric and amperometry methods with short response time, as inexpensive instruments and can be used for direct onsite measurement. Wan et al. [24, 25] reported a fast and sensitive approach for simultaneous determination of lead and copper based on a commercial screen-printed gold electrode with gold nanoparticle modification. Nano band electrode array was also used to determine lead and copper. However, those are unsuitable for in situ monitoring, due to complicated measurement procedure. Planar Interdigital (ID) capacitive sensors have finger-like, parallel, in-plane electrode geometry with repetitive patterns on a solid substrate. Planar geometry-based electrodes have non-destructive, single side access, improved penetration depth and low cost which have been used for different biological and chemical sensing applications [26, 27]. This type of sensors has been used for different applications, such as quality monitoring of dairy, meat, and leather products [28, 29].

The Internet of Things (IoT) enables any physical objects to communicate through the internet and transfer data to a specific server for further processing. It requires pervasive computing, smart sensors, embedded devices, communication technologies, internet protocols and applications. IoT-enabled WSNs help to connect more objects for monitoring purposes to build up smart cities, smart industries, smart agriculture, etc. In Chap. 7, a low-cost interdigital FR4 based capacitive sensor is developed for the measurement of nitrate. The developed sensor has been characterised and calibrated against standards to measure nitrate components. Standard UV-spectrometry method is used to validate the experiment. A smart sensing node has been developed which is suitable for an IoT-enabled WSN. The sensing node collects water from stream or lake and measures the nitrate concentration. The collected water can be discharged automatically to stream or lake again. A solar panel is used for energy harvesting and provides the system with the necessary energy for monitoring.

Chapter 8 is concluded the book chapters and some of the recommendations are provided as a part of future work.

References

1. E. Berner, R. Berner, *The Global Water Cycle* (Prentice Hall, Upper Saddle River, 1987)
2. J. Gaillard, *Lecture on Nitrogen Cycle* (1995)
3. W.H. Organization, in Nitrate and nitrite in drinking-water. http://www.who.int/water_sanitation_health/dwq/chemicals/nitrate-nitrite-background-jan17.pdf. Accessed 26 April 2018
4. H.H. Comly, Cyanosis in infants caused by nitrates in well water. JAMA **129**(2), 112–116 (1945)
5. C.J. Johnson et al., Fatal outcome of methemoglobinemia in an infant. JAMA **257**(20), 2796–2797 (1987)
6. P.F. Swann, The toxicology of nitrate, nitrite and n-nitroso compounds. J. Sci. Food Agric. **26**(11), 1761–1770 (1975)
7. C.S. Bruning-Fann, J. Kaneene, The effects of nitrate, nitrite and N-nitroso compounds on human health: a review. Vet. Hum. Toxicol. **35**(6), 521–538 (1993)
8. O. G. Licenase, in The water supply (water quality) regulations 2016. Accessed 26 April 2018

9. J. Davis, K.J. McKeegan, M.F. Cardosi, D.H. Vaughan, Evaluation of phenolic assays for the detection of nitrite. Talanta **50**(1), 103–112 (1999)
10. M. Friedberg, M. Hinsdale, Z. Shihabi, Analysis of nitrate in biological fluids by capillary electrophoresis. J. Chromatogr. A **781**(1–2), 491–496 (1997)
11. P.N. Bories, E. Scherman, L. Dziedzic, Analysis of nitrite and nitrate in biological fluids by capillary electrophoresis. Clin. Biochem. **32**(1), 9–14 (1999)
12. G. Ellis, I. Adatia, M. Yazdanpanah, S.K. Makela, Nitrite and nitrate analyses: a clinical biochemistry perspective. Clin. Biochem. **31**(4), 195–220 (1998)
13. A. Azmi, A.A. Azman, S. Ibrahim, M.A.M. Yunus, Techniques in advancing the capabilities of various nitrate detection methods: a review. Int. J. Smart Sens. Intell. Syst. **10**(2) (2017)
14. P. Brimblecombe, D. Stedman, Historical evidence for a dramatic increase in the nitrate component of acid rain. Nature **298**(5873), 460 (1982)
15. G.R. Hallberg, Nitrate in ground water in the United States, in *Developments in Agricultural and Managed Forest Ecology*, vol. 21 (Elsevier, Amsterdam, 1989), pp. 35–74
16. G. Hallberg, D. Keeney, in *Nitrate, Regional Groundwater Quality*, ed. by W.J. Alley (Van Nostrand Reinhold, New York, 1993)
17. D. Behm, Ill Waters: the fouling of Wisconsin's lakes and streams (special report). Milwaukee J. **2** (1989)
18. M.A.M. Yunus, S. Ibrahim, W.A.H. Altowayti, G.P.San, S.C. Mukhopadhyay, Selective membrane for detecting nitrate based on planar electromagnetic sensors array, in *10th Asian Control Conference (ASCC)* (IEEE, 2015), pp. 1–6
19. C. Wardak, Solid contact nitrate ion-selective electrode based on ionic liquid with stable and reproducible potential. Electroanalysis **26**(4), 864–872 (2014)
20. R.K. Mahajan, R. Kaur, H. Miyake, H. Tsukube, Zn (II) complex-based potentiometric sensors for selective determination of nitrate anion. Anal. Chim. Acta **584**(1), 89–94 (2007)
21. T. Kjær, L.H. Larsen, N.-P. Revsbech, Sensitivity control of ion-selectivebiosensors by electrophoretically mediated analyte transport. Anal. Chim. Acta **391**, 57–63 (1999)
22. J. Aylott, D. Richardson, D. Russell, Optical biosensing of nitrate ions using asol–gel immobilized nitrate reductase. Analyst **122**, 77–80 (1997)
23. P.T. Charles, P.R. Gauger, C.H. Patterson, A.W. Kusterbeck, On-siteimmunoanalysis of nitrate and nitroaromatic compounds in groundwater. Environ. Sci. Technol. **34**, 4641–4650 (2000)
24. H. Wan, Q. Sun, H. Li, F. Sun, N. Hu, P. Wang, Screen-printed gold electrodewith gold nanoparticles modification for simultaneous electrochemicaldetermination of lead and copper. Sens. Actuators B: Chem. **209**, 336–342 (2015)
25. H. Wan, Q. Sun, H. Li, F. Sun, N. Hu, P. Wang, Design of a miniaturizedmultisensor chip with nanoband electrode array and light addressablepotentiometric sensor for ion sensing. Anal. Methods **7**, 9190–9197 (2015)
26. C. Berggren, B. Bjarnason, G. Johansson, Capacitive biosensors. Electroanalysis **13**, 173–180 (2001)
27. W. Limbut, M. Hedström, P. Thavarungkul, P. Kanatharana, B. Mattiasson, Capacitive biosensor for detection of endotoxin. Anal. Bioanal. Chem. **389**, 517–525 (2007)
28. S.C. Mukhopadhyay, C.P. Gooneratne, G.S. Gupta, S.N. Demidenko, A low-costsensing system for quality monitoring of dairy products. IEEE Trans. Instrum. Meas. **55**, 1331–1338 (2006)
29. S. Mukhopadhyay, S.D. Choudhury, T. Allsop, V. Kasturi, G. Norris, Assessmentof pelt quality in leather making using a novel non-invasive sensing approach. J. Biochem. Biophys. Methods **70**, 809–815 (2008)

Chapter 2
Literature Review

2.1 Introduction

There are various detection methods are available for nitrate detection in water. Some of them are laboratory-based methods, some can perform in situ measurement. Some of the detection methods are conventional, some of them are non-conventional methods. All the detection methods are some advantages and drawbacks. The price of the detection methods are also important and needs to consider during measurement. This chapter discusses all the available methods and their characteristics in terms of advantages, limit of detections, detection methods, drawbacks, cost and important features.

2.2 Detection Methodologies

There are two methods for nitrate detection: direct and indirect methods. Indirect methods are expensive, as costly instrumentation is required for accurate measurement. They are also complicated, require chemical reagents, produce much chemical waste and need the expertise to make the measurements. On the other hand, direct methods are accurate and economical. A significant disadvantage of using a direct method is measurement errors due to interference from other contaminants. Both methods are discussed in the subsequent sections.

2.3 Electrochemical Detection

The electrochemical detection of nitrate can be divided into many different categories depending on their different sensing methods. The system can convert the nitrate ions into an impedance, potential difference or current, and they can be grouped into

Md E. E. Alahi and S. C. Mukhopadhyay, *Smart Nitrate Sensor*, Smart Sensors, Measurement and Instrumentation 35, https://doi.org/10.1007/978-3-030-20095-4_2

impedimetric, potentiometric and voltammetric, respectively. In an electrochemical cell, various electrodes can be used for nitrate detection, such as copper, silver, platinum, gold, glassy-carbon electrode (GCE), graphite-epoxy, chitosan/bentonite, graphene, and many more. The LOD, the sensitivity of the sensor, and its reusability depend on the material of the electrodes and the sensing method.

This method is used widely due to its simple operation, good sensitivity to nitrate ions in water, easy miniaturisation, and low power consumption. However, the conventional electrochemical cells are large and not suitable as a portable device. Some research [1–3] is reported where nitrate detection was done through a miniature electrochemical cell. One of the significant reasons to develop a low-cost portable device is to monitor the nitrate concentration continuously in aqueous media. Hence, the developed analytical system needs to be highly miniaturised and sensitive enough for nitrate detection over a long period. Therefore, the reusability of any proposed analytical system is extremely important. Different electrochemical detection methods are discussed in the following sub-sections.

2.4 Potentiometric Detection

The conventional potentiometric system consists of two half-cells which contain an electrode immersed in a solution of ions, where the activities of the electrode's potential are determined. An inert electrolyte in a salt bridge, such as potassium chloride (KCl), connects the two half-cells. Porous frits are fixed to the ends of the salt bridge. It allows the electrolyte's ions to move between the half-cells and the salt bridge. Figure 2.1 represents the architecture of a potentiometry system.

This detection method was first introduced in late 1976 for detection of nitrate ions in water. It does not require any additional chemical reagent and is considered a direct detection method. In the last few years, this method has been utilised to improve the selectivity and detection limits of nitrate detection by enhancing the lackings of the method to improve the performance of the system, such as the integration of ion-selective electrodes (ISE). The ISE method is useful to determine the free-ion concentration directly in water. The advantages of using ISE are low cost, non-

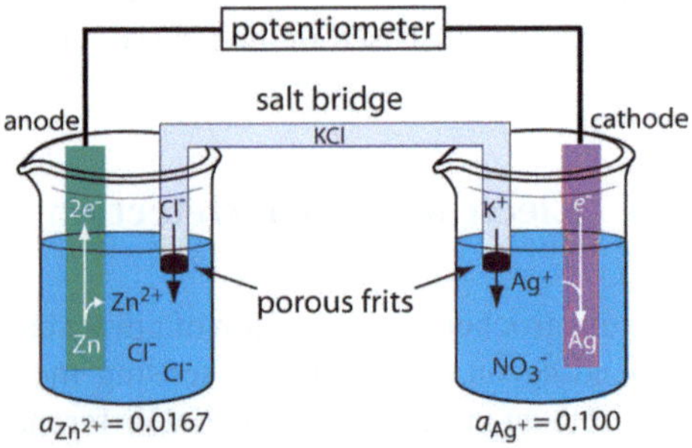

Fig. 2.1 Architecture of potentiometry system

destruction of the sample, portable device, and no requirement for pre-treatment of the samples.

In 1976, Hassan et al. [4] reported organic nitrate and nitramine determination based on the reaction of a mercury sulphuric-acid mixture. The advanced potentiometric method was simple, selective and accurate for rapid detection at the micro and sub-micron level. The detection limit was 1–50 μM with a precision of $\pm 0.2\%$. Mendezo et al. [5] developed a nanobiocomposite as ISE to determine nitrate ions in water. The nanocomposite is based on intercalation of chitosan in bentonite, and the detection limit was 20–800 mM. Mahajan et al. [6] developed a polymeric membrane employing two Zn(II) complexes coordinated by neutral tetradentate ligands, N, N′-ethylene-bis (N-methyl-(S)-alanine methylamide) and N, N′-ethylene-bis (N-methyl-(S)-alanine dimethylamide), utilised as anion-selective carriers. The blending of these new Zn (II) complexes with dioctyl sebacate offered a high sensing selectivity for nitrate ions in water. The detection range was near-Nernstian slopes in the wide linear concentration range of 50 uM to 100 mM. Li [7] and Nunez [8] developed the potentiometric sensor and used an Artificial Neural Network (ANN) to determine the nitrate contamination level in the water. Bendikov and Harmon [9] have reported doped polypyrrole, which is a selective membrane in an ISE electrode for nitrate detection in water. The doped polypyrrole is a highly conductive polymer material, which is widely used due to its high conductivity and stability. Zhang et al. [10] used this doped polypyrrole as a sensitive polymer membrane for a potentiometric system for nitrate detection in water. The polypyrrole improves the selectivity with a simple recipe procedure and has low toxicity compared to polyvinyl chloride (PVC) which is a conventional membrane [10]. Moreover, this reported study also demonstrated that carbon-nanostructure based materials between the membrane and the substrate layers are useful to prevent water formation during the potentiometric measurement. Wardak et al. [11] developed trihexyltetradecylphosphonium chloride (THTDPIC) which is an active polymeric membrane and imposes on the PVC membrane sensitivity by reducing electrical resistance.

Most of the potentiometric sensors are bulky due to the internal reference electrode and the reference electrolyte solution with a true-liquid or liquid polymeric membrane. Thus, a micro-fabricated polymeric sensor [12, 13] was introduced to reduce that problem with these potentiometric sensors. The advantages of using micro-fabrication are small size, simple design, reduced cost, and mass production. A lot of different materials have been introduced, such as screen-printed thick film, silicon-based transducer, metal-printed flexible film, to produce micro-scale potentiometric sensors. They have shown an excellent response to detecting nitrate ions in water. Figure 2.2 is an example of the response from a potentiometric sensor and Fig. 2.3 is the layout and design of a potentiometric sensor. Table 2.1 summarises the characteristics of various potentiometric sensors.

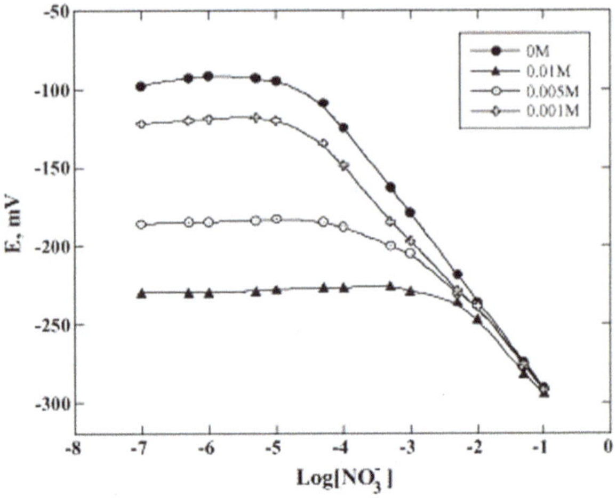

Fig. 2.2 Potentiometric response of an electrode at varying concentrations [6]

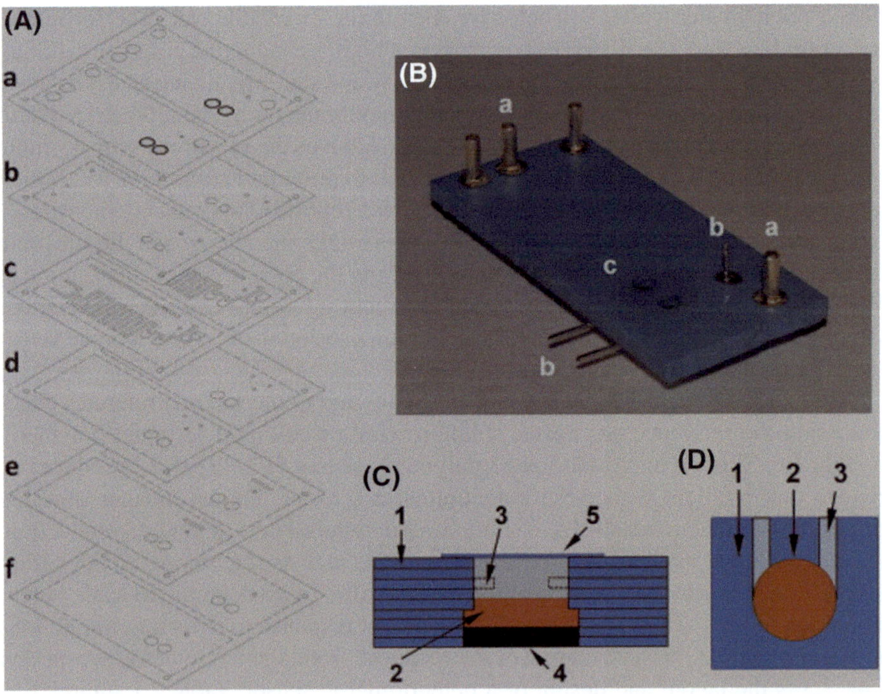

Fig. 2.3 **a** Layout of the biparametric prototype; **b** picture of the final constructed device; **c** front view of the detection-chamber scheme; **d** Top view of the detection-chamber scheme [12]

Table 2.1 Characteristics of various potentiometric sensors

Electrode and membrane	Reference electrode	Limit of Detection (LOD)	Nernstian slope (mV/decade)	Response time (s)	Application	Characteristics	References
Silver bis (bathophenanthroline) nitrate in a plasticized PVC	Ag/AgCl	0.05 μgm/L	−55.1 ± 0.1	<15	Determination of nitrate in industrial wastewater, fertilizers and pharmaceuticals.	Fast response, high sensitivity, long-term stability and good selectivity	[13]
Graphite-epoxy and PVC	Ag/AgCl	4.6 μmol/L	−68.2	Not reported	Photoelectrocatalytic treatment of nitrate monitoring	Simple operation does not require any preparative or pre-treatment stage	[8]
Nitrate polymeric	Ag/AgCl	9.56 mg/L	−59.5	Not reported	In the water treatment process plant	Small in size, simultaneous and on-line detection	[12]
Glassy carbon electrode and polypyrrole with nitrate doped	Ag/AgCl	10–4.8 mol/L	−55.1 ± 0.1	85	Monitoring of soil micronutrient	Rapid response, inexpensive cost, simple operation	[10]
Chitosan/bentonite and graphite-epoxy	Ag/AgCl	2×10^{-4} M	−54.6	Not reported	Determination of nitrate in water	Low cost, easy to use	[5]

(continued)

Table 2.1 (continued)

Electrode and membrane	Reference electrode	Limit of Detection (LOD)	Nernstian slope (mV/decade)	Response time (s)	Application	Characteristics	References
Trihexyltetradecylphosphonium chloride, plasticizer and PVC membrane	Ag/AgCl	2.8 μmol/L	−60.1	5–10	Determination of nitrate in water	Simple, fast and cheap	[11]
Pencil lead (graphite) and doped polypyrrole	Silver wire	$5 \pm 1 \times 10^{-5}$ M	−54	Not reported	Nitrate determination from tap water, stream or lake water	Simple, easy to prepare and low cost	[9]
PVC, N,N′-ethylene-bis (N-methyl-(S)-alanine methylamide) and N,N′-ethylene-bis (N-methyl-(S)-alanine dimethylamide), Zn (II)	Ag/AgCl	1.0×10^{-3} M	−55.16	25	Determination of nitrate in water	Fast response, reusability for 3 months	[6]

2.5 Amperometric Detection

Amperometry is an electrochemical method where the potential of the sensing electrode is controlled through instrumentation, and the current is recorded as the analytical signal, appearing because of oxidation/reduction. The applied potential is constant, and the resulting current is measured as a function of time. This technique, first introduced in 1976 as high-performance liquid chromatography (HPLC) [14], and a conventional benchtop was introduced in 1987 [15]. During the amperometry process, the magnitude of the generated current in any given sample is determined by the number of molecules and can be calculated using Faraday's law:

$$i_t = \frac{dQ}{dt} = \eta F \frac{dN}{dt} \tag{2.1}$$

where, i_t is the current generated at the sensing surface electrode during time t, Q is the charge at the sensing surface, η is the number of moles, N is the number of moles of the analyte during oxidation/reduction, and F is the Faraday constant (96487 C mol^{-1}). The detection limit can be at attomole or femtomole levels [16, 17]. Therefore, this method is useful to determine nitrates in the water (Fig. 2.4).

Carpenter et al. [19] reported an amperometric method to determine the nitrate in water. The nitrate concentration range was 0.1–1 mM, and interferences from other similar ions in water can be avoided. Zhang [20] reported Polypyrrole (PPy)-nanowire modified electrodes based on graphite electrodes for nitrate detection. The detection limit was 1.52 μM, and the sensitivity reaches 336.28 mA/M cm^2. The results showed that it has significant effects on the morphology of PPy nanowires and the current density of nitrates during the electrochemical preparation of the modified electrodes. Other amperometric detectors [21–30] are reported to determine nitrate in water. The design of the sensing electrodes is essential to increase the sensitivity of the detection method.

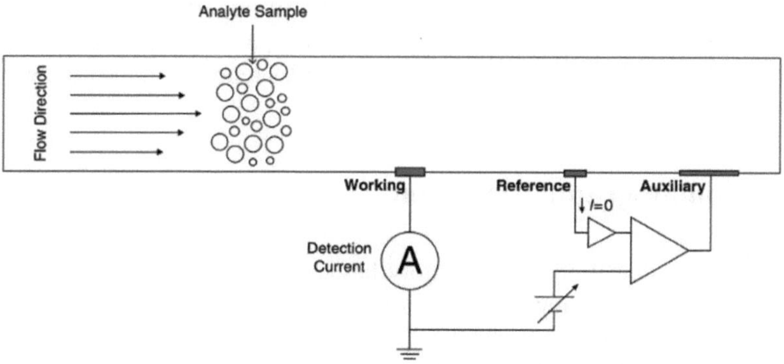

Fig. 2.4 Schematic diagram of amperometric detection cell for lab-on-a-chip application [18]

2.6 Voltammetric Detection

Cyclic voltammetry (CV) is a powerful and popular electrochemical detection method that measures the reduction and oxidation process of any molecular species. It is usually performed with three electrodes named reference electrode, working electrode and counter electrode. The working electrode carries out the electrochemical event of interest that changes the potential applied these to give the desired potential at the reference electrode. The reference electrode is used as a reference point to measure the potential of the other electrodes. During proving the potential to the working electrode, current begins to flow, and the auxiliary electrode completes the circuit to continue the flow of electrons due to the reduction or oxidation process.

In 1977, Bidini et al. [31] reported a voltammetric method to determine the nitrate in irrigation water samples. Copper and cadmium were electrochemically deposited on the pyrolytic graphite electrode which was the working electrode. The linear detection range was from 1 μM to 1 mM. Davenport et al. [32] reported a voltammetric determination of nitrate and nitrite using a rotating cadmium disk. Krista et al. [33] reported a system where the electrodes were prepared from a mixture of silver, graphite powder and methacrylate resin. The detection range of nitrate ion was up to 31 mg/L, and LOD was 7 mg/L. The arrangement showed good reproducibility, but a computer-controlled system was necessary to get those results. Shahriar et al. [34] reported an in situ copper- based electrode with a low detection limit. Differential pulse voltammetry (DPV) was applied for simultaneous nitrate determination in river water. Neuhold et al. and Mareček et al. [35, 36] reported the voltammetric determination of nitrate in drinking water. The detection limit was 0.5–60 μg/mL, and a carbon-paste electrode was used as the working electrode. A differential pulse voltammetric method [37] has been reported to determine the nitrate in natural water. The detection limit was 2.8 μM, and the linear detection range was 2.8 μM–80 μM. Copper-plated glassy carbon was used, with a sensitivity of 0.9683 A. L/mol. A square-wave voltammetric method [38] has been reported with similar copper- plated glassy carbon electrodes. The linear working range of determination of nitrate was 0.61–50 μM and LOD was 0.18 μM. Jang et al. [39] reported a new electrochemical sensing platform with self-assembly nanobeads-packed (nBP) hetero-columns. Figure 2.5 illustrates the sensing platform which was reported in the article.

2.7 Chromatography Detection

Chromatography detection is a laboratory-based technique used for different kinds of chemical mixture separation. An anion or cation of the analyte sample can be extracted when performing a chemical reaction with an eluent. It is an essential technique for various chemical and biological analyses. There are available several chromatography techniques, such as ion chromatography [40], high-performance liquid chromatography [41], and ultra-performance liquid chromatography [42].

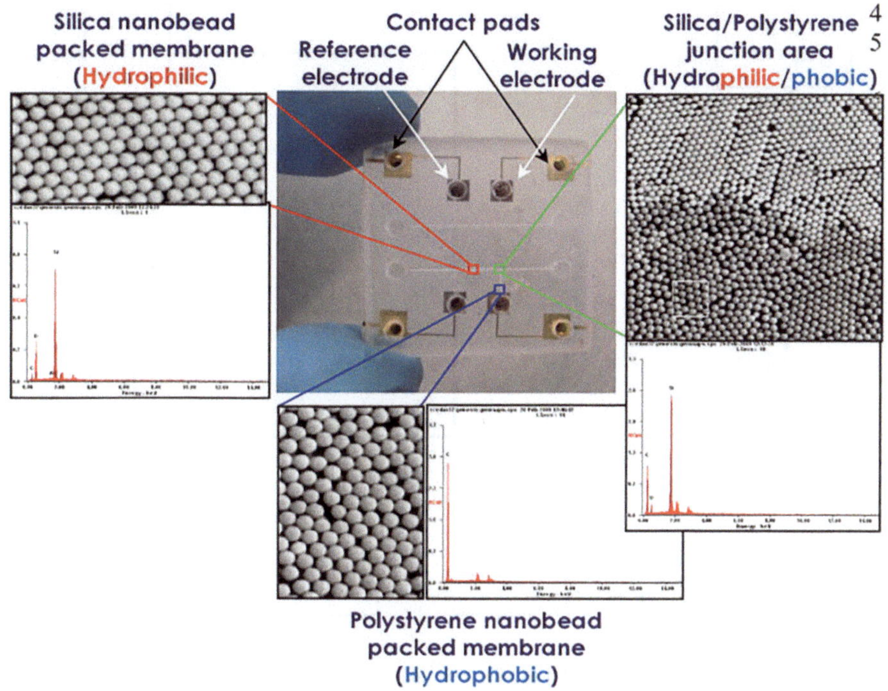

Fig. 2.5 Photograph of the dual ion-selective lab chip with self-assembly nBP columns [39]

Ion chromatography is widely used for nitrate detection in water [43–49]. An ion-chromatography system consists of solvents, solvent degasser, pump, injector, pre-column heat exchanger, guard column, post-column heat exchanger, electrolytic suppressor, and detector. The system is expensive, bulky, and not suitable for in situ nitrate measurement.

This detection method also includes UV spectrometry, fluorimetric, electron capture, and mass spectroscopy [40, 41, 43, 44, 50–52]. UV detection is a popular method due to its simplicity. The low absorbance of water maintains a low background, which reduces the complications for direct detection of nitrate ions in water. However, rising of the high background UV absorbance is always avoidable which might come using the aromatic eluent during the detection of nitrate (Fig. 2.6).

A novel HPLC method is also available to detect nitrate with improved sensitivity [41]. It is simple, easy to handle and selective towards nitrate ions in water, and is based on a photochemical reaction and ion-exchange separation. Niedzielski et al. [43] reported a technique which can be useful to determine nitrate, nitrite and ammonium ions in water. Zuo et al. [52] developed a simple, fast, accurate HPLC method to detect nitrate and nitrite ions in lake water. The method produced good sensitivity, a high detection range, and a good LOD. Table 2.2 summarises the type of Chromatography methods with their characteristics.

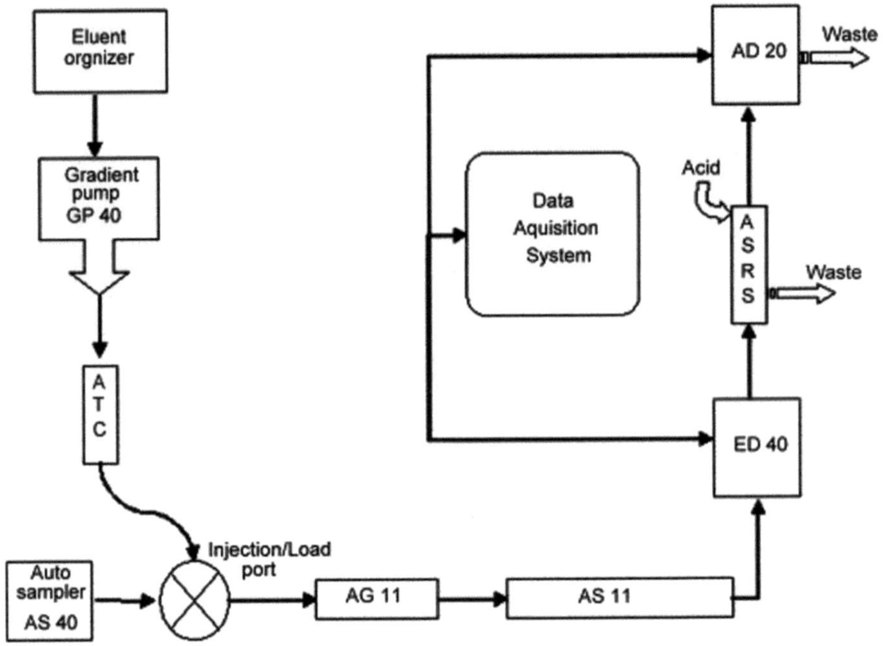

Fig. 2.6 Schematic representation of the ion-chromatography setup used for simultaneous determination [44]

2.8 Biosensors

The biosensing method is a direct method to detect nitrates in the water. In biosensing systems, biological materials are employed with a detection system and a signal-conditioning circuit to measure the concentration of the targeted ions in a sampling solution. The sampling or analyte solution is exposed directly to the biosensor and the sensing material to measure the concentration of the sample. The target ions interact with the biological material to provide the necessary information. The interaction process needs to be converted to an electrical signal such as voltage, current or impedance to measure from the sensing system. The conversion of the sensing signal depends on the biosensing method. Useful articles have reported [53–56] on biosensors and differentiated them based on the nature of the transducer. Biological components such as DNA, enzymes, immunological systems, receptor proteins, and whole cells can be used as recognition units. A transduction component that may be acoustic, chemical, electrochemical, microbalance, optical, or piezoelectric is required. They should have high sensitivity, specificity, and the ability to work in a wide range of matrices. They are also useful if they operate remotely to make in situ measurements.

For the last two decades, nitrate biosensors have been developed due to the easy availability of appropriate enzymes and their use in a substrate as a material. Zeng

Table 2.2 The types of chromatography with eluent and LOD

Chromatography type	Detector	Eluent	LOD	Application	References
Ion chromatography	Ion exchange and diode-array detection	NaCl	0.05 mg/L	Detection of nitrate in rainwater	[43]
Ion chromatography	low-capacity anion exchange column with amperometric and absorbance	NaOH	6 μg/L	Detection of nitrate in water	[44]
HPLC	Ion-exchange separation, online photochemical reaction, and luminol chemiluminescence	Borate buffer (pH 10)	2×10^{-8} M	Detection of nitrate in river, rainfall, pond, tap and commercial mineral water	[41]
HPLC	UV light absorption	Tetrabutylammonium hydroxide, Na_2HPO_4 (pH 3.9), and Acetonitrile	5 μg/L	Detection of nitrate detection	[52]
Liquid chromatography	Fluorescence detection	Toluene and NaOH	0.3 pg/L	Biological, food and environmental samples	[50]

et al. [57] reported a fluorescence-based fibre-optic bio-sensor to trace various contaminants in seawater. A protein molecule was used as a recognition unit. Xuejiang et al. [58] reported a fast, sensitive and stable conductometric enzyme biosensor for the determination of nitrate in water. A methyl viologen mediator modified the electrodes of the biosensor, mixed with nitrate reductase (NR). The developed biosensor had a quick response and reached 95% of the constant conductance value in 15 s. The calibration range is 0.02–0.25 mM with a detection limit of 0.005 mM. The lifetime of the sensor was two weeks. Cosnier et al. [59] used NR by entrapping in a laponite clay gel which was cross-linked by glutaraldehyde. The developed sensor was useful to detect nitrate in low concentration. The use of biosensors for the measurement of nitrate in water could enhance the sensitivity for in situ remote monitoring. Can et al. [3] reported an amperometric nitrate biosensor where the film was produced from Polypyrrole (PPy)/Carbon nanotubes (CNTs). The sensitivity of the developed sensor was 300 nA/mM, and the detection range was 0.44–1.45 mM. The biosensor gave a higher response than the standard methods of nitrate measurement. Zhang et al. [60] reported a conductometric biosensor. Table 2.3 tabulates the type of biological materials, the detection systems, and the limit of detection (LOD).

2.9 Flow-Injection Analysis

Flow-based methods have been used for nitrate measurement for various applications. This detection method is desirable in the laboratory and is an excellent alternative to other traditional methods. It has high-throughput analysis, requires low volumes of reagent and sample, is low-cost and easy to operate. Hence, the reported work [63–72] used the flow-injection analysis method and developed a procedure to determine the nitrate concentration in water.

Generally, flow-injection methods consist of four different processes: distribution, reduction, detection, and detection. The sample is carried on to the next process with a constant flow rate for the distribution process. The pump of the flow-injection methods allows the supplied sample to proceed endlessly within a specific period. Second is a reduction process which works by converting the nitrate ions into active nitrite ions by elevating the nitrate ions inside the sample. In the detection process, a compound is formed due to the reaction of a reducing agent with the nitrate ions produced in the earlier process. Then, the responsibility of the detection system is to detect nitrate ions. The most common detection agents are Griess-illosvay, acidic hydrogen peroxide, sulfanilamide and N-(1-naphthyl) ethylenediamine dihydrochloride (NED).

In the flow-injection process, several reduction reagents have been used for nitrate detection. A zinc column, a cadmium column, vanadium (III), titanium (III) chloride and hydrazine sulfate are the most-common reduction reagents. It is imperative to select the correct reduction reagent; else, it will take an unnecessary amount of time to complete the injection process. Hydrazine et al. [70] reported a process which requires a longer reaction time for the reduction process, and hazardous reagents can be associated as a by-product which can contribute to pollution.

Different types of detectors are involved with the numerous flow-injection methods, and they have been reported earlier for nitrate detection. Among the different types of detection systems, spectrophotometric and chemiluminescence are preferred methods for nitrate detection. Absorbance is the output of the detector which is proportional to the sample concentrations. Lambert's law explains this behaviour where absorbance is proportional to the concentration of sample nitrate. The chemical reagent, pH of the solution, flow rate of the sample carrier, sample volume and coil length of the reaction are related to getting the maximum output of detection. All these parameters must be controlled to achieve accurate results, sample throughput, peak shape of the output. Therefore, all those parameters are optimized for the analysis. Figure 2.7 shows a schematic diagram of the essential components of a flow-injection analyser.

Table 2.3 Types of biological materials, detection systems, limit to detect

Detection method	Biological material	LOD	Application	Characteristics	References
Conductometric	Methyl viologen mediator with NR	0.005 mM	Detection of nitrate in wastewater and river water	Response time was 15 s, low detection limit. It had good operational and thermal stability. Life of the sensor not more than two weeks	[58]
Amperometric	CNT/PPy film electrode with NR	0.17 mM	Measurement of nitrate in water	The developed sensor had better performance than standard analysis. Response time was 20 s and the first 10 days was better	[3]
Cyclic voltammetry	NaR–SOD1–CNT–PPy–Pt	200 nM	Nitrate detection in saliva and blood	Extremely sensitive with large sample size required for the measurements	[61]
Amperometric	*Escherichia coli* with NR	0.1 mM	Detection of nitrate in drinking water	Longer stability and cheap method to prepare the biosensor	[62]
Cyclic voltammetry	Laponite clay gel, glutaraldehyde with NR	7 µM	Determination of nitrate	Low detection limit	[59]

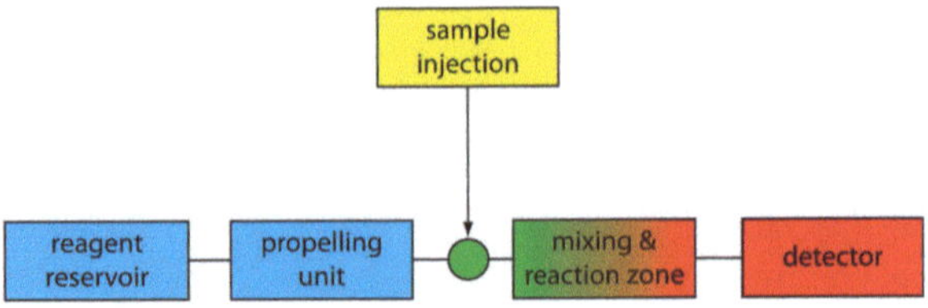

Fig. 2.7 Schematic diagram of necessary components of a simple flow-injection analyser

Yaqoob et al. [65] reported on luminal chemiluminescence, where the efficiency depends on the pH, and the pH helps in reduction of nitrate; increasing the pH changes the absorbance, or the detector's response, significantly to develop the azo dye formation which increases the efficiency of spectrophotometric detection [64]. The flow rate is the vital parameter, and it is controlled for the sample, carrier and reagent solutions. The significance of the flow rate is reported in [66], and it is essential for time-consuming reactions.

Furthermore, it can enhance the sensitivity and sample throughput for nitrate ions [69]. It is important to find the optimum flow rate which helps to stabilise the baseline on the detection area and improve the output such as intensity or absorbance. Table 2.4 summarises the various characteristics of flow-injection analysis.

2.10 Electromagnetic Sensors

2.10.1 Planar Electromagnetic Sensor

Nowadays, much research on impedance-based sensors has been performed. They are perfect for measuring physical properties and suitable for direct measurement. A Planar Electromagnetic Sensor Array (PESA) is one of those sensors which measure physical properties in terms of impedance. A sample has different properties, such as conductivity, dielectric properties, and permeability which can be estimated to measure the nitrate in water samples [80–87]. Enhancing the sensor's sensitivity is essential and can be done by optimising the sensor design and configuration. PESA is convenient to use as an in situ measurement which is highly reliable, with fast response, and low cost.

Yunus et al. [81] reported a planar electromagnetic sensor for nitrate detection in an aqueous medium. The performance of the sensor was measured through two arrangements: series and parallel connections. The study showed that the series connection provided better performance in terms of the sensor's sensitivity to detect the nitrate. Therefore, it was recommended to use the series connection for electromagnetic sensors. The sensitivity of the sensors also depends on the material of the electrodes. The reported work used gold electrodes due to the attraction of nitrate towards gold. It is also essential to use a high-dielectric substrate to increase the

Table 2.4 Flow-injection methods with various characteristics

System	Reduction agent and detection agent	LOD	Detection system	Application	References
Flow analysis	Zinc granules and N-(1-naphthyl) ethylene diamine (Griess reagent)	1.3 μg/L	Spectrophotometric	Nitrate measurement in marine and estuarine waters	[73]
Flow analysis	Enzymatic assay reagent and Griess-Ilosvay	0.73 μg/L	Spectrophotometric	Determination of nitrate and nitrite	[74]
Automated parametric	Cadmium and Modified Griessi-llosvay	0.0207 mg N/L	Spectrophotometric	Determination of nitrate, nitrite, and sulfate in drinking water	[75]
Flow injection	Vanadium (III) Chloride and Sulfanilamide and N-(1-naphthyl)ethylenediamine dihydrochloride (NED)	0.1 μM	Spectrophotometric	Determination of nitrite and nitrate	[76]
Flow injection	Cadmium and sulfanilamide and N- (1- naphthyl) ethylenediamine dihydrochloride	50 μgN/L	Spectrophotometric	Nitrate and nitrite determination in natural waters	[77]
Flow injection	Cadmium and sulfanilamide and N-(1-naphthyl) ethylenediami dihydrochloride	0.013 μgN/L	Spectrophotometric	Determination of nitrate, nitrite, and ammonium in soil	[78]
Flow injection	Cadmium and hydrogen peroxide	0.02 μg N/L	Luminolchemi luminescence	Determination of nitrate and nitrite in freshwater	[79]

penetration depth of the electric field, to develop a high-sensitivity sensor. Three configurations are available to design the electrodes: parallel, star, and delta.

Nor et al. [85] reported a sensor array with a thin substrate on a Printed Circuit Board (PCB), made using a conventional PCB fabrication technique. The designed sensor consisted of several coils or loops of electrodes, spiral or square in shape. The distance between the coils is essential to improve the sensitivity of the sensor. Wang et al. [88] reported the design of a sensor where the impedance of the sensor increased due to increasing the distance between the two electrodes and decreasing their area.

2.10.2 *Planar Interdigital Sensor*

The operating principle of the planar interdigital sensor is similar to that of a parallel-plate capacitor, where the electrodes provide a single measurement to the material under test (MUT) [89]. The electric field lines generated from the positive electrode and bulge through the MUT in their path towards the negative electrode. The sensor behaves as a capacitor, and the impedance depends on the physical properties of the MUT. Therefore, it is possible to know the physical properties of the material by knowing the impedance behaviour of the interdigital sensor (Fig. 2.8).

Some work is reported [90–96] on an interdigital sensor for nitrate measurement to provide a low-cost sensing system. The measurement was done through Electrochemical Impedance Spectroscopy (EIS), where a small alternating signal is provided to the positive electrode and an output signal is taken from the negative electrode. Impedance measurement is done through an LCR meter. The length, width and pitch gaps are essential to increase the sensitivity of the sensor. The material of the electrodes and the sensing area are also essential to improve the sensitivity of an interdigital sensor. Alahi et al. [91] reported a smart sensor for the detection of nitrate in water. The detection limit is 0.01–0.5 mg/L, and temperature compensation is added to improve the sensitivity. Gold electrodes were used for excitation and sensing on the silicon substrate. The pitch gap of the electrodes was 25 μm and the dimensions of the sensing surface 2.5 mm × 2.5 mm. The developed system can

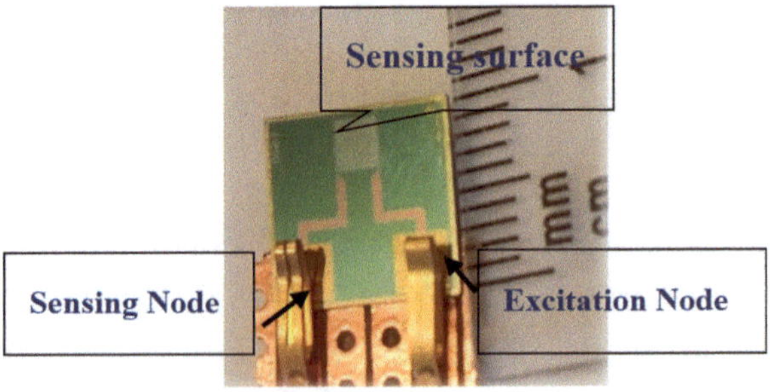

Fig. 2.8 MEMS-based interdigital sensor [90]

measure the nitrate in water and transfer the measured data to the cloud for real-time monitoring. Another reported work of Alahi et al. [92] explained the principle of the particular material for nitrate detection. The material was developed by an imprinting polymer technique and had a detection range from 1 to 10 mg/L. The developed material was used as a coating material on the interdigital sensor to detect nitrate in lake, stream, river and canal water. The calibration curve (Fig. 2.9) was developed in terms of reactance, and later it was used to measure unknown nitrate samples.

A Graphene-based electrode (Fig. 2.10) was used [94] to determine the nitrate in water, where the detection limit was 1-70 mg/L. The graphene was printed on Kapton tape to improve the sensitivity for the measurement. Temperature compensation was also proposed to improve the sensitivity of the sensor. The sensor was robust, low-cost, easy to use and useful for nitrate measurement without any selectivity. The same group [97] reported on nitrate measurement. The interdigital sensor was based on FR4, and the Internet of Things (IoT) and so is beneficial to detect nitrate in real time. The system was useful to measure water from creek, stream, lake and get the data remotely from the sampling location. Table 2.5 summarises the various characteristics of planar interdigital sensors (Fig. 2.11).

2.11 Fibre-Optic Sensor

In the last three decades, an optical sensor has been developed to determine various parameters in water. It consists of three essential parts: a source, an optical fibre and a photodetector to detect the optical signal. The optical sensor can be chosen from two categories: intrinsic and extrinsic sensors. The intrinsic sensors require continuous, consistent light sources to permit a phase-modulation technique. The modulation

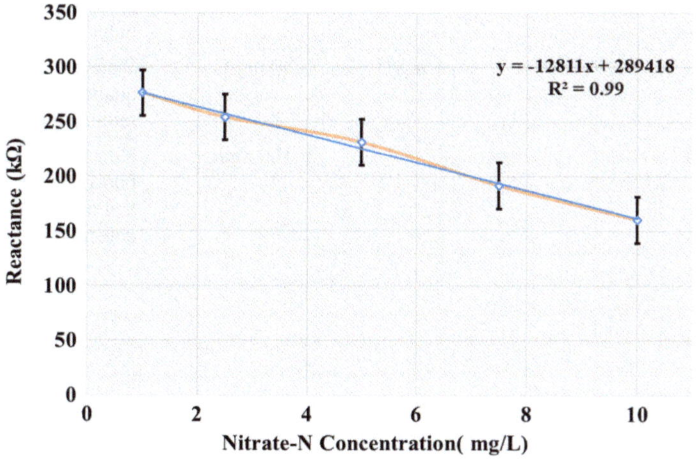

Fig. 2.9 Calibration standard for nitrate-N measurement [92]

Fig. 2.10 Graphene-based interdigital sensor [94]

Table 2.5 Various planar interdigital sensor with different characteristics

Type of sensor	Sensor's response	Limit of Detection (LOD)	Linear range	Application	References
MEMS sensor	−115,543 Ω/0.1 ppm	0.01 mg/L	0.01–0.5 mg/L	Smart sensing system for agricultural industry	[91]
Graphene Sensor	−667.97 Ω/ppm	1 mg/L	1–70 mg/L	IoT-enabled smart sensing system	[94]
MEMS-based Coated Sensor	−12,811 Ω/ppm	1.06 mg/L	1–10 mg/L	Smart sensing system for water quality measurement	[92]
FR4-based sensor	−51.76 Ω/ppm	1 mg/L	1–40 mg/L	Water quality monitoring	[97]
FR4-based coated sensor	–	–	5–100 mg/L	Nitrate, phosphate and nickel detection	[99]

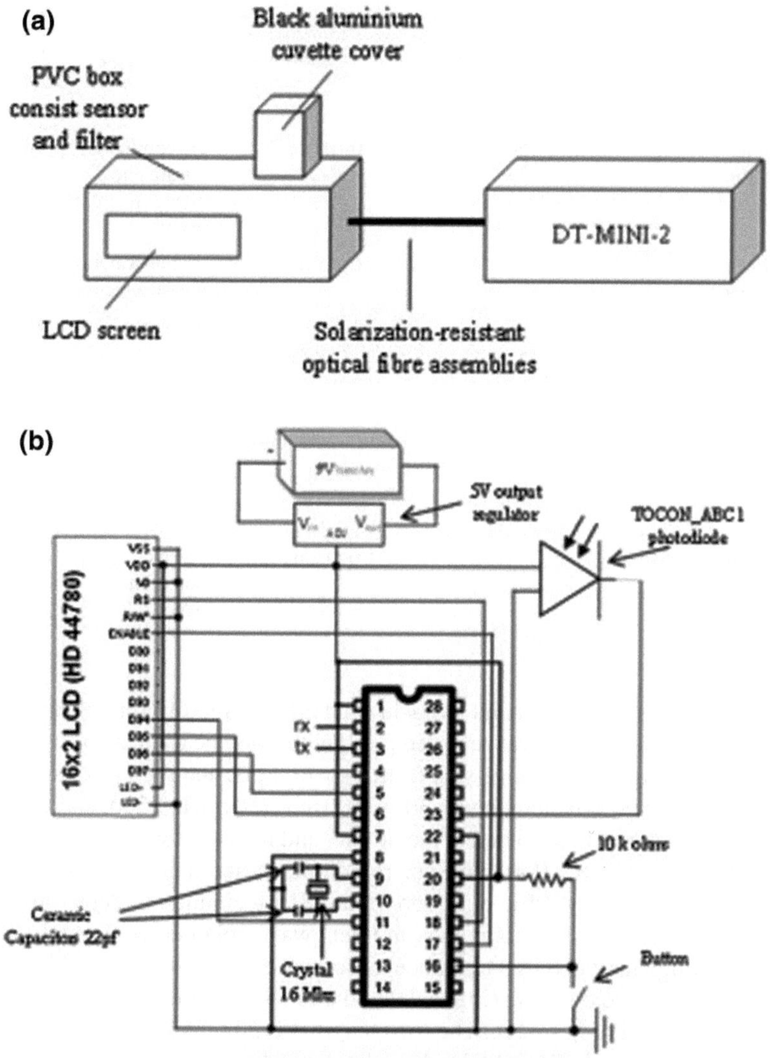

Fig. 2.11 **a** Setup of the fiber-optic sensor and **b** schematic of the sensor with microcontroller system [98]

occurs inside the optical fibre, and only a single-mode fibre is used for the design of the sensor. Fibre is used as a transmitting channel for an external optic sensor. It is widely used in remote sensing due to its low power requirement and small size. It is also used to detect nitrate and nitrite [100] in water. Lauth's violet-triacetyl cellulose membrane was used to detect nitrate concentration using absorption spectrophotometry, and the detection range is 10.12–1012 ng/mL. Mahieuxe [101] reported a modified fibre-optic sensor using fluorescence emission to detect nitrate in water. It

Table 2.6 Different optical-fibre sensors and their characteristics

Type of sensor	Analytical wavelength (nm)	Limit of detection (LOD)	Linear range	Application	References
Ultraviolet optical fibre sensor	302	0.0017 mg/L	0–50 mg/L	Detection of nitrate and nitrite in water	[98]
Ultraviolet Spectrometry	240	0.4 μmol/L	0–45 μmol/L	Ocean-water measurement	[104]
Coating-based optical fibre sensor	300–1100	–	1–70 mg/L	Nitrate measurement in drinking water	[102]
Removal of cladding from optical fiber	350–2500	–	0–2.50 mg/L	Nitrate measurement in water	[103]
Optical-fiber chemical sensor	515	–	0–0.2 μmol/L	Nitrate detection in water	[101]

presents a bathochromic shift and hypochromic effect with different nitrate concentrations. An external sensor was reported [102] with Lophine used as a sensitive layer on the fibre. The detection range was from 1 to 70 mg/L and the response time was 20 ms. The detection wavelength was 300–1100 nm. 350 nm to 2500 nm [103] wavelengths have been reported to detect the nitrate, and the range was 0–2.50 mg/L. An ASD FieldSpec 3 Hi-Res Portable Spectroradiometer and halogen lamp were used as light source. Recently, Moo et al. reported [98] a technique to detect nitrate and nitrite in water. The spectroscopic measurement was implemented through different techniques which included transmittance, absorbance, and reflectance. All measurements were presented by using Channel 0 of the Jaz Spectrometer. Tungsten halogen light was used as the light source. The wavelength was 302–356 nm with the detection range from 0 to 50 mg/L. The obtained results showed a good linear relationship with the minimum effect of interference from other ions. Jhonson et al. [104] reported an in situ ultraviolet spectrophotometry method to do long-term nitrate measurements in ocean water. The nitrate concentration is measured by the absorption of the UV light produced by a deuterium light source and guided through fibres to a reflection probe. The in situ ultraviolet spectrometer is controlled by a microcontroller, which also contained a global positioning system and satellite antenna. The linear range of detection of nitrate is 0–10 ppm. This sensor has been used in various applications [105–110] due to the improved sensitivity. Table 2.6 represents the various characteristics of optical-fibre sensors.

2.12 Commercial Sensors

Hach is well-known as suppliers of analytical instruments and chemical reagents for the laboratory-based purpose, especially for water quality and other liquid solutions. Take IntelliCal ISENO3181 Nitrate Ion Selective Electrode (ISE) for nitrate detection as an example, it is designed by the solid-state PVC membrane with epoxy, and solid gel ion exchange eliminates the frequency of replacing membrane. The ISE can only be used periodically in the labour field and is not suited to continual measurement. It can detect the nitrate-N concentration from 0.1 to 14,000 mg/L. It has an integrated temperature sensor which can measure the temperature range from 0 to 50 °C. The price is at $2100. Hach Nitratax sc tank sensor is being used extensively internationally and in New Zealand. It uses the UV absorption measurement with reagent-free technique. There are three different models with the different detection range: 0.1–100 mg/L by Nitratax plus sc; 1.0–20 mg/L by Nitratax exo sc; and 0.5–20 mg/L by Nitratax clear sc. These models cost around $40 K–60 K.

"S::can" is Austria-based Company which provides varied product range for water and environmental monitoring. All "s::can" instruments are operated followed by the "plug & measure" principle, so all of them are ready to use with pre-calibrated works. The "spectro::lyser" UV monitors can be used to NO_3-N detection. It is measured based on the UV-Vis spectrometry with the nitrate detection range from 0 to 20 mg/L. This instrument is also being used extensively internationally and in New Zealand. However, the whole set of the system is costly (around $60–70 K).

Hanna Instruments is focused on developing electro-analytical instrumentation. They provide the products like nitrate portable/benchtop photometer and nitrate ion selective electrode. For example, Hanna HI96728 is a portable checker to detect nitrate in freshwater based on the colourimetric method. A particular Tungsten lamp is used as the light source, and silicon photocell with narrow band interference filter at the wavelength of 525 nm used to detect the light. It uses the cadmium reduction method to determine the nitrate-nitrogen concentration range from 0.0 to 30.0 mg/L. The price is at around NZD 350.

Xylem is a world leader in providing small instruments for water technology. It has a broad range of products under different brands. YSI is one of Xylem brand, which provides environmental monitoring products. IQ SensorNet 182 is a modular water quality terminal which allowed connecting additional sensors such as 6884 Nitrate ISE sensor and it provides a continuously measurement of water quality parameters like pH, Dissolved oxygen, temperature, conductivity, ammonium, nitrate, potassium, TOC (total organic carbon), COD (chemical oxygen demand), DOC (dissolved organic carbon share of TOC), BOD (biochemical oxygen demand) and SAC (spectral absorption coefficient). It has an easy-to-read digital display and wireless connection via radio transmission with a range of 100 m. However, when it uses the nitrate sensor such as NitraLyt, the function of real-time monitoring is disabled. Another Xylem brand—WTW, it provides online measurement system—TresCon Analyzer which can monitor Ammonium, Nitrate, and Nitrite continuously. A continuous water sample supply with low solids contents is required for operating this

Table 2.7 Summary of commercially available sensors

Model	Method of detection	Detection range (nitrate)
IntelliCal ISENO3181	Ion selective electrode	0.1–14,000 mg/L
Nitratax plus sc	UV-photometric	0.1–100 mg/L
S::can	UV-spectrometric	0–20 mg/L
Hanna HI96728	Colorimetric	0–30 mg/L
YSI 6884 Nitrate Sensor	Ion selective electrode	0.0–200.0 mg/L
TresCon Analyzer	UV-spectrophotometric	0.1–60 mg/L
UV Nitrate Monitor AV455	Ion exchange	0–100 mg/L
ChemScan 6101 Process Analyzer	UV-spectrometric	0.5–20 mg/L

system. For nitrate detection, the UV light is absorbed by the nitrate ions and determines the nitrate concentration at a wavelength of 254 nm.

ABB is a multinational corporation which mainly designs and manufactures power and automation products, including flow measurement, gas, and liquid analyser and environmental monitoring systems. For example, UV Nitrate Monitor AV455 provides a continuous analysis without chemical reagents. It can detect the nitrate concentration at the range of 0–100 mg/L at a wavelength of 220 nm. It requires low-maintenance and simple calibration with auto cleaning for the optical component. It uses de-nitrification as a process to reduce the nitrate concentrations for nitrate monitoring.

ASA Analytics Inc. specialises in the manufacture of the automatic chemical analysis system. ChemScan 6101 Process Analyzer can monitor Ammonia, Phosphate, Nitrate-Nitrite, and Phosphate in surface water based on the UV-spectrometric method. The light is absorbed by nitrate ions and into 256 wavelengths of 200–450 nm. The detection range of nitrate is from 0.5 to 20 mg/L (Table 2.7).

Table 2.1 summarises the commercially available nitrate detection product/system in the market. Hanna HI 96728 is a portable device, but it still needs to collect the sample and require the reagent for measurement of nitrate. The YSI IQ SensorNet 182 terminal provides continuous monitoring, but it calls for an additional sensor for nitrate measurement.

2.13 Challenges and Future Opportunities

Even though much work has been done with WFSs, there are still some issues that need to be dealt with. Researchers are trying 24/7 to develop sensors with better performance, regarding sensitivity and sustainability, than the existing ones and more comfortable fabrication and implementation, for ubiquitous monitoring. The massive amount of data generated by the sensors wearable system causes a difficulty to handle and store them. Also, it becomes a tedious job for the system to filter out significant data from the massive database for future analysis. Due to the enormous amount of monitored data, there needs to be a proper security system to curb any mishandling and misuse of the received data. Time-varying traffic is another issue raised during data transmission from different sensor nodes in a real-time topological system. This causes a delay in data reception in the monitoring unit, thus decreasing the efficiency of the system. Also, some of the significant data might get lost due to the high traffic generated by AAL applications. The data transmission for a central coordinator system in Wireless Sensor Networks (WSNs) should be appropriately handled to minimise traffic and the loss of data. The connectivity and interoperability of the embedded system should be significant, to minimise the power and data loss. From a patient's point of view, the person should not face any discomfort to wear the WFS. There should not be any breach of privacy for the patient from the monitoring. The embedded system attached to the sensor should not loosely attach to the body or clothes worn by the person, who could alter the data depending on movements and the surrounding environment. There is also a risk of thermal effects of the attached sensors from the tissues of the patient. Many factors decide these thermal effects [111]. The number of sensors used in the embedded system should be kept as small as possible. The location of the sensor is also important. The positioning of the sensor on the arm will have more thermal effects than its positioning on the chest. The operating frequency of the sensor and network protocol should be as low as possible. Power consumption by WFSs is another significant issue that needs to be addressed. Sensors like SHIMMER, Telos with low power consumption should be considered for monitoring purposes to reduce the overall power consumed by the WFS. The continuous supply of power to the system is another challenge that needs to be addressed for future systems. The system should be designed for on-node processing and reduce the effects of motion artifact and distributed interference.

There is a prominent future for flexible electronics in wearable systems based on its market values [112]. The market value of printed and flexible electronics is estimated to be over 75 billion USD by 2025 [113]. There is a substantial opportunity to use these flexible systems for monitoring health parameters. The estimated cost of WFSs by 2020 is more than 3 billion USD [114], and over 40 billion USD with more than 240 million annual unit shipments by 2025 [115]. The challenge for the companies is to design the systems to decrease in the overall fabrication cost. One way to achieve this is to use cheap, safe and biocompatible materials for the designs. Flexible, one of the UK-based companies, has predicted a rise in organic electronics among a WFSs [116]. With the growing interest of consumers, the companies should design systems which would serve the people, to meet not only the application purposes but also their economic condition. The systems should be made cost-effective so that it can address the broader community in the society.

2.14 Internet of Things (IoT)

Internet of Things (IoT) is the general concept of things that are readable, recognisable, locatable, addressable, and controllable via the Internet. Things communicate through the internet via Radio Frequency Identification (RFID), Wireless Local Area Network (WLAN), Wide Area Network (WAN) and other means. In the last few years, it has been a gaining appreciation with the revolution of advanced wireless technology. The IoT can be defined as a global network structure, linking physical and virtual objects, things, and devices through intelligent objects, sensors, and communication and actuation capabilities [117, 118].

Different types of sensors are becoming available for sensing physical events and transmitting sensor data via wireless communications. Sensor networks are becoming a significant component of network architecture and applications. However, more current sensor networks are designed and deployed to provide only a particular application. The sensor-nodes/platforms where the sensors are capable of connecting with embedded electronics, real-time sensing capability, and access to any other nodes are the most crucial part of the future IoT [119–122]. Figure 2.12 shows the architecture of IoT of three-tier communication, i.e., the local, regional and global tiers. The first tier consists of wireless sensing nodes/platforms. The next tier is the regional base station tier where the first tier acquires raw sensor data or intelligence and sends them to this tier through light-weight protocol stack [i.e., Zigbee, 6LoWPAN, RFID]. The base station works as the medium between the sensor-node tier and the clouds which are also a global tier and routes the raw sensing data to another node or a remote server [119, 123]. Appropriate tools and approaches are needed for testing and managing an application on real hardware on a large scale to design a robust application. In early IoT research, the availability of smart devices is limited. Only recent advances in technology have increased their availability at a lower cost. Although the experiments were mainly on a small scale and conducted in a research laboratory, they allowed for an opportunity to enhance understanding of the impact and limitations of protocols and design choices on performances of real hardware [124].

Another essential term of IoT application is cloud computing. It is a model for enabling convenient, on-demand network access to a shared pool of configurable computing resources that can be rapidly provisioned and released with minimal management effort or service provider interaction [125]. In an IoT based cloud computing platform, instead of having local servers for collecting and managing information coming from WSN networks, remotely located servers (implemented utilising virtual and physical machines) are dynamically provisioned and changed, according to the actual application. So IoT based application should have the ability to access remotely through internet [126].

Fig. 2.12 The architecture of three-tier IoT [119]

2.15 Existing Sensor Nodes for IoT

The solutions to overcome the problems and challenges relevant to the wireless sensor network (WSN) research have developed widely in recent years. A variety of sensor-nodes have provided a solid basis for experimentation and moving the IoT to a globally networked infrastructure. This opens up new challenges that demand new capabilities and features from suitable sensor-nodes. In the following section, we have reviewed the currently available sensor-nodes for IoT experiments in a range of different applications [127–129].

Most of the new sensor systems have been built using early-stage wireless sensor network platforms such as TelosB [130], Imote2 [131], MICAz [132] and TinyOS software framework. The sensor network hardware platforms are low-power embedded microcontroller systems with some embedded sensors and analogue I/O ports to connect. A set of software components also needs to be developed for IoT application, such as real-time OS, sensor/hardware drivers, networking protocols, and application-specific sensing and processing algorithms. MoteLab2 was one of the first and longest lasting testbeds which are used by Harvard University to set up an experimental wireless sensor network [133]. INDRIYA3 is a three-dimensional wireless sensor network testbed developed at the National University of Singapore, following the same design as MoteLab [134]. SHIMMER was used for biomedical health care purposes, and it was one of the extremely flexible sensor platforms. It could expand seamlessly to meet the various biomedical research project requirements. The sensor platform had massive research potential, as explained in [135, 136]. Imote2 was used to analyze the health of a civil structure that could localise the damage at different solutions. It selectively activates nodes in the damaged region to achieve fine-grained localisation [137, 138]. CoSMoS is another platform that was used to monitor the health and safety of construction workers. More work has been done on environmental monitoring compared to other applications. In most of the cases, they have used Arduino and Raspberry Pi which are robust, low-cost platforms for sensor-nodes [139]. Sensor4PRI is another platform where they have used an extensive network based on Arduino MEGA [140]. It is also a powerful platform to analyse the sensor data. The IEEE 802.15.4 standard has the physical and medium access control layers for low data-rate wireless personal area networks. ZigBee is a low-cost, low-power, wireless mesh networking standard. It is offered as commercial-off-the-shelf (COTS) modules for developing different prototyping of wireless sensing and actuation systems. It is also used in the different applications of IoT [141–146].

From Table 2.8, it is seen that in most cases, ZigBee has been used as a network protocol to communicate with the server or to transfer the data for different application purposes. Its low power consumption limits the transmission distances to 10-100 m, depending on power output and environmental characteristics. In certain applications, Zigbee devices can transmit data over long distances by passing data through a mesh network. There are specific modules available in the market. From all of those, it is proposed to use the WiFi card as a part of this sensor-node.

Table 2.8 Different sensor platforms and their noteworthy, drawbacks

Applications	Sensor platform, CPU	Noteworthy	Drawbacks
Biomedical research for health care [135]	SHIMMER™, TI MSP430	Extremely flexible sensor platform. Actively used for Kinematic, physiological and ambient sensing. TinyOS and Chipcon CC2420 [147] are used	Working at research level only
Distributed structural health monitoring [117, 137]	mote2, Intel PXA271	Flexibility based structural engineering method that can localize damage at different resolutions. It selectively activates nodes in the damaged region to achive fine grained localization. Tiny OS and Chipcon CC2420 are used	The node selection process is not automatic
Programmable pervasive spaces [148]	Atlas, ATmega128L	A single pervasive space can cover much geographically dispersed area. Extremely large or densely packed environment could result in poor performance. Tiny OS is used. Chipcon CC2420	In an extremely large or densely packed environment many services could result in poor performances

(continued)

Table 2.8 (continued)

Applications	Sensor platform, CPU	Noteworthy	Drawbacks
Smart parking system & precision irrigation system [123]	Z50, ATmega328P	Low cost and energy efficient for low data rate monitoring. TinyOS is used. Hoperf RFM12b (ISM Band FSK Transceiver)	Lower flash memory and lower data rate (115.2 kbps)
WSN with optical fiber pH sensor [149]	Novel Hybrid WSN platform, TI MSP430	It showed that an optical fibre pH sensor module could be successfully created and integrated into a WSN module. Communication protocol and GUI software have developed also. CC2530F256 (SoC Transceiver, Zigbee)	The mesh network does not have 'self-healing' capability
Health and safety of a construction worker [139]	KoSMoS (TelosB), TI MSP430	Developed facility management for workers. TinyOS is used. IEEE 802.15.4 compliant RF transceiver	The sensor-node has a reliability problem. Continuous monitoring is not possible because of limited battery power
TABLE I. Environmental monitoring [150]	Arduino, Raspberry Pi, ATmega328	Low cost, compact, scalable, easily customized, easy to deploy, easy to maintain. Humidity and temperature are measured. X-Bee of Zig-bee family radio module is used	The base station has limited storage space considering the application
TABLE II. Experimental WSN network [133]	MoteLab, ATmega128L	It was an experimental wireless sensor network deployed at Harvard University. One of the first and longest lasting testbeds. Chipcon CC1000 (RF transceiver)	Earlier sensor-node which is not useful for complex mesh network. 128 Kbytes RAM memories are available
WSN network [134]	INDRIYA (TelosB), TI MSP430	Wireless sensor network developed inNational University of Singapore. Chipcon CC2420 (ZigBee-ready RF Transceiver)	RAM memory is small which would not be enough to store the entire raw sensor data
Protection of railway infrastructure [140]	Sensor4PRI (Arduino), ATmega2560	Railway infrastructure protection and sensor platform designed to be integrated into a slab track system. X-Bee radio module is used	Large amount of energy is consumed while transmitting data

References

1. E. Andreoli et al., Electrochemical conversion of copper-based hierarchical micro/nanostructures to copper metal nanoparticles and their testing in nitrate sensing. Electroanalysis **23**(9), 2164–2173 (2011)
2. S. Aravamudhan, S. Bhansali, Development of micro-fluidic nitrate-selective sensor based on doped-polypyrrole nanowires. Sens. Actuators B Chem. **132**(2), 623–630 (2008)
3. F. Can, S.K. Ozoner, P. Ergenekon, E. Erhan, Amperometric nitrate biosensor based on carbon nanotube/polypyrrole/nitrate reductase biofilm electrode. Mater. Sci. Eng., C **32**(1), 18–23 (2012)
4. S.S. Hassan, Ion-selective electrodes in organic functional group analysis: Microdetermination of nitrates and nitramines with use of the iodide electrode. Talanta **23**(10), 738–740 (1976)
5. M.O. Mendoza, E.P. Ortega, O.A. de Fuentes, Y. Prokhorov, J.G.L. Barcenas, Chitosan/bentonite nanocomposite: preliminary studies of its potentiometric response to nitrate ions in water, in *2014 IEEE 9th Ibero-American Congress on Sensors (IBERSENSOR)* (IEEE, 2014); pp. 1–4
6. R.K. Mahajan, R. Kaur, H. Miyake, H. Tsukube, Zn (II) complex-based potentiometric sensors for selective determination of nitrate anion. Anal. Chim. Acta **584**(1), 89–94 (2007)
7. C. Li, L. Li, Prediction of nitrate and chlorine in soil using ion selective electrode, in *World Automation Congress (WAC)* (IEEE, 2010),pp. 231–234
8. L. Nuñez, X. Cetó, M.I. Pividori, M.V.B. Zanoni, M. Del Valle, Development and application of an electronic tongue for detection and monitoring of nitrate, nitrite and ammonium levels in waters. Microchem. J. **110**, 273–279 (2013)
9. T.A. Bendikov, T.C. Harmon, A sensitive nitrate ion-selective electrode from a pencil lead. An analytical laboratory experiment. J. Chem. Educ. **82**(3), 439 (2005)
10. L. Zhang, M. Zhang, H. Ren, P. Pu, P. Kong, H. Zhao, Comparative investigation on soil nitrate-nitrogen and available potassium measurement capability by using solid-state and PVC ISE. Comput. Electron. Agric. **112**, 83–91 (2015)
11. C. Wardak, Solid Contact Nitrate Ion-Selective Electrode Based on Ionic Liquid with Stable and Reproducible Potential. Electroanalysis **26**(4), 864–872 (2014)
12. A. Calvo-López, E. Arasa-Puig, M. Puyol, J.M. Casalta, J. Alonso-Chamarro, Biparametric potentiometric analytical microsystem for nitrate and potassium monitoring in water recycling processes for manned space missions. Anal. Chim. Acta **804**, 190–196 (2013)
13. S.S. Hassan, H. Sayour, S.S. Al-Mehrezi, A novel planar miniaturized potentiometric sensor for flow injection analysis of nitrates in wastewaters, fertilizers and pharmaceuticals. Anal. Chim. Acta **581**(1), 13–18 (2007)
14. P.T. Kissinger, T.H. Ridgway, Small-amplitude controlled-potential techniques, in *Laboratory Techniques in Electroanalytical Chemistry, Revised and Expanded* (1996), p. 141
15. R.A. Wallingford, A.G. Ewing, Capillary zone electrophoresis with electrochemical detection. Anal. Chem. **59**(14), 1762–1766 (1987)
16. S. Sloss, A.G. Ewing, Improved method for end-column amperometric detection for capillary electrophoresis. Anal. Chem. **65**(5), 577–581 (1993)
17. J. Wang, *Analytical Electrochemistry* (Wiley, London, 2006)
18. T.J. Roussel, D.J. Jackson, R.P. Baldwin, R.S. Keynton, Amperometric Techniques, in *Encyclopedia of Microfluidics and Nanofluidics* (2013), pp. 1–11, 2013
19. N.G. Carpenter, D. Pletcher, Amperometric method for the determination of nitrate in water. Anal. Chim. Acta **317**(1–3), 287–293 (1995)
20. X.-L. Zhang, J.-X. Wang, Z. Wang, S.-C. Wang, Improvement of amperometric sensor used for determination of nitrate with polypyrrole nanowires modified electrode. Sensors **5**(12), 580–593 (2005)
21. J.R.C. da Rocha, L. Angnes, M. Bertotti, K. Araki, H.E. Toma, Amperometric detection of nitrite and nitrate at tetraruthenated porphyrin-modified electrodes in a continuous-flow assembly. Anal. Chim. Acta **452**(1), 23–28 (2002)

22. J.E. Newbery, M.P.L. de Haddad, Amperometric determination of nitrite by oxidation at a glassy carbon electrode. Analyst **110**(1), 81–82 (1985)
23. M.A. Stanley et al., Comparison of the analytical capabilities of an amperometric and an optical sensor for the determination of nitrate in river and well water. Anal. Chim. Acta **299**(1), 81–90 (1994)
24. J.C. Gamboa, R.C. Pena, T.R. Paixão, M. Bertotti, A renewable copper electrode as an amperometric flow detector for nitrate determination in mineral water and soft drink samples. Talanta **80**(2), 581–585 (2009)
25. A. Hulanicki, W. Matuszewski, M. Trojanowicz, Flow-injection determination of nitrite and nitrate with biamperometric detection at two platinum wire electrodes. Anal. Chim. Acta **194**, 119–127 (1987)
26. G.A. Sherwood, D.C. Johnson, A chromatographic determination of nitrate with amperometric detection at a copperized cadmium electrode. Anal. Chim. Acta **129**, 101–111 (1981)
27. S.A. Glazier, E.R. Campbell, W.H. Campbell, Construction and characterization of nitrate reductase-based amperometric electrode and nitrate assay of fertilizers and drinking water. Anal. Chem. **70**(8), 1511–1515 (1998)
28. A.Y. Chamsi, A.G. Fogg, Oxidative flow injection amperometric determination of nitrite at an electrochemically pre-treated glassy carbon electrode. Analyst **113**(11), 1723–1727 (1988)
29. M. Bertotti, D. Pletcher, Amperometric determination of nitrite via reaction with iodide using microelectrodes. Anal. Chim. Acta **337**(1), 49–55 (1997)
30. M.A. Alawi, Determination of nitrate and nitrite in water with HPLC and amperometric detection. Fresenius' J. Anal. Chem. **313**(3), 239–240 (1982)
31. M.E. Bodini, D.T. Sawyer, Voltammetric determination of nitrate ion at parts-per-billion levels. Anal. Chem. **49**(3), 485–489 (1977)
32. R.J. Davenport, D.C. Johnson, Voltammetric determination of nitrate and nitrite ions using a rotating cadmium disk electrode. Anal. Chem. **45**(11), 1979–1980 (1973)
33. J. Krista, M. Kopanica, L. Novotný, Voltammetric determination of nitrates using silver electrodes. Electroanal. Int. J. Fundamental Practical Aspects Electroanal. **12**(3), 199–204 (2000)
34. S.M. Shariar, T. Hinoue, Simultaneous voltammetric determination of nitrate and nitrite ions using a copper electrode pretreated by dissolution/redeposition. Anal. Sci. **26**(11), 1173–1179 (2010)
35. V. Mareček, H. Jänchenová, Z. Samec, M. Březina, Voltammetric determination of nitrate, perchlorate and iodide at a hanging electrolyte drop electrode. Anal. Chim. Acta **185**, 359–362 (1986)
36. C. Neuhold, K. Kalcher, W. Diewald, X. Cai, G. Raber, Voltammetric determination of nitrate with a modified carbon paste electrode. Electroanalysis **6**(3), 227–236 (1994)
37. A.O. Solak, P. Gülser, E. Gökm, F. Gökmeşe, A new differential pulse voltammetric method for the determination of nitrate at a copper plated glassy carbon electrode. Microchim. Acta **134**(1–2), 77–82 (2000)
38. A. Osman Solak, P. Çekirdek, Square wave voltammetric determination of nitrate at a freshly copper plated glassy carbon electrode. Anal. Lett. **38**(2), 271–280 (2005)
39. A. Jang, Z. Zou, K.K. Lee, C.H. Ahn, P.L. Bishop, Potentiometric and voltammetric polymer lab chip sensors for determination of nitrate, pH and Cd (II) in water. Talanta **83**(1), 1–8 (2010)
40. C. Lopez-Moreno, I.V. Perez, A.M. Urbano, Development and validation of an ionic chromatography method for the determination of nitrate, nitrite and chloride in meat. Food Chem. **194**, 687–694 (2016)
41. H. Kodamatani, S. Yamazaki, K. Saito, T. Tomiyasu, Y. Komatsu, Selective determination method for measurement of nitrite and nitrate in water samples using high-performance liquid chromatography with post-column photochemical reaction and chemiluminescence detection. J. Chromatogr. A **1216**(15), 3163–3167 (2009)
42. M.R. Siddiqui, S.M. Wabaidur, Z.A. ALOthman, M. Rafiquee, Rapid and sensitive method for analysis of nitrate in meat samples using ultra performance liquid chromatography–mass spectrometry. Spectrochimica Acta Part A Mol. Biomol. Spectroscopy **151**, 861–866 (2015)

43. P. Niedzielski, I. Kurzyca, J. Siepak, A new tool for inorganic nitrogen speciation study: Simultaneous determination of ammonium ion, nitrite and nitrate by ion chromatography with post-column ammonium derivatization by Nessler reagent and diode-array detection in rain water samples. Anal. Chim. Acta **577**(2), 220–224 (2006)
44. K. Tirumalesh, Simultaneous determination of bromide and nitrate in contaminated waters by ion chromatography using amperometry and absorbance detectors. Talanta **74**(5), 1428–1434 (2008)
45. M. Tabatabai, W. Dick, Simultaneous determination of nitrate, chloride, sulfate, and phosphate in natural waters by ion chromatography 1. J. Environ. Qual. **12**(2), 209–213 (1983)
46. J.A. Morales, L.S. de Graterol, J. Mesa, Determination of chloride, sulfate and nitrate in groundwater samples by ion chromatography. J. Chromatogr. A **884**(1–2), 185–190 (2000)
47. I. Dahllöf, O. Svensson, C. Torstensson, Optimising the determination of nitrate and phosphate in sea water with ion chromatography using experimental design. J. Chromatogr. A **771**(1–2), 163–168 (1997)
48. E. Kapinus, I. Revelsky, V. Ulogov, Y.A. Lyalikov, Simultaneous determination of fluoride, chloride, nitrite, bromide, nitrate, phosphate and sulfate in aqueous solutions at 10–9 to 10–8% level by ion chromatography. J. Chromatogr. B **800**(1–2), 321–323 (2004)
49. M. Neal, C. Neal, H. Wickham, S. Harman, Determination of bromide, chloride, fluoride, nitrate and sulphate by ion chromatography: comparisons of methodologies for rainfall, cloud water and river waters at the Plynlimon catchments of mid-Wales. Hydrol. Earth Syst. Sci. **11**(1), 294–300 (2007)
50. M. Akyüz, Ş. Ata, Determination of low level nitrite and nitrate in biological, food and environmental samples by gas chromatography–mass spectrometry and liquid chromatography with fluorescence detection. Talanta **79**(3), 900–904 (2009)
51. Y. Li, J.S. Whitaker, C.L. McCarty, Reversed-phase liquid chromatography/electrospray ionization/mass spectrometry with isotope dilution for the analysis of nitrate and nitrite in water. J. Chromatogr. A **1218**(3), 476–483 (2011)
52. Y. Zuo, C. Wang, T. Van, Simultaneous determination of nitrite and nitrate in dew, rain, snow and lake water samples by ion-pair high-performance liquid chromatography. Talanta **70**(2), 281–285 (2006)
53. S. Rodriguez-Mozaz, M.J.L. de Alda, D. Barceló, Biosensors as useful tools for environmental analysis and monitoring. Anal. Bioanal. Chem. **386**(4), 1025–1041 (2006)
54. S. Rodriguez-Mozaz, M.J.L. de Alda, D. Barceló, Fast and simultaneous monitoring of organic pollutants in a drinking water treatment plant by a multi-analyte biosensor followed by LC–MS validation. Talanta **69**(2), 377–384 (2006)
55. B. Roig, I. Bazin, S. Bayle, D. Habauzit, J. Chopineau, Biomolecular recognition systems for water monitoring, in *Rapid Chemical and Biological Techniques for Water Monitoring* (2009), pp. 175–195
56. M. Farré, L. Kantiani, S. Pérez, D. Barceló, Sensors and biosensors in support of EU Directives. TrAC Trends Anal. Chem. **28**(2), 170–185 (2009)
57. H.-H. Zeng, R.B. Thompson, B.P. Maliwal, G.R. Fones, J.W. Moffett, C.A. Fierke, Real-time determination of picomolar free Cu (II) in seawater using a fluorescence-based fiber optic biosensor. Anal. Chem. **75**(24), 6807–6812 (2003)
58. W. Xuejiang et al., Conductometric nitrate biosensor based on methyl viologen/Nafion®/nitrate reductase interdigitated electrodes. Talanta **69**(2), 450–455 (2006)
59. S. Cosnier, S. Da Silva, D. Shan, K. Gorgy, Electrochemical nitrate biosensor based on poly (pyrrole–viologen) film–nitrate reductase–clay composite. Bioelectrochemistry **74**(1), 47–51 (2008)
60. Z. Zhang et al., A novel nitrite biosensor based on conductometric electrode modified with cytochrome c nitrite reductase composite membrane. Biosens. Bioelectron. **24**(6), 1574–1579 (2009)
61. T. Madasamy, M. Pandiaraj, M. Balamurugan, K. Bhargava, N.K. Sethy, C. Karunakaran, Copper, zinc superoxide dismutase and nitrate reductase coimmobilized bienzymatic biosensor for the simultaneous determination of nitrite and nitrate. Biosens. Bioelectron. **52**, 209–215 (2014)

62. D. Albanese, M. Di Matteo, C. Alessio, Screen printed biosensors for detection of nitrates in drinking water,in *Computer Aided Chemical Engineering*, vol. 28 (Elsevier, Amsterdam, 2010), pp. 283–288
63. A. Ayala, L. Leal, L. Ferrer, V. Cerdà, Multiparametric automated system for sulfate, nitrite and nitrate monitoring in drinking water and wastewater based on sequential injection analysis. Microchem. J. **100**, 55–60 (2012)
64. M. Yaqoob, A. Nabi, P.J. Worsfold, Determination of nitrite and nitrate in natural waters using flow injection with spectrophotometric detection. J. Chem. Soc. Pakistan **34**(3) (2013)
65. M. Yaqoob, B. Folgado Biot, A. Nabi, P.J. Worsfold, Determination of nitrate and nitrite in freshwaters using flow-injection with luminol chemiluminescence detection. Luminescence **27**(5), 419–425 (2012)
66. S. Wang, K. Lin, N. Chen, D. Yuan, J. Ma, Automated determination of nitrate plus nitrite in aqueous samples with flow injection analysis using vanadium (III) chloride as reductant. Talanta **146**, 744–748 (2016)
67. C.L. Pasquali, A. Gallego-Picó, P.F. Hernando, M. Velasco, J.D. Alegría, Two rapid and sensitive automated methods for the determination of nitrite and nitrate in soil samples. Microchem. J. **94**(1), 79–82 (2010)
68. C.L. Pasquali, P.F. Hernando, J.D. Alegria, Spectrophotometric simultaneous determination of nitrite, nitrate and ammonium in soils by flow injection analysis. Analytica chimica Acta **600**(1–2), 177-182 (2007)
69. S. Feng, M. Zhang, Y. Huang, D. Yuan, Y. Zhu, Simultaneous determination of nanomolar nitrite and nitrate in seawater using reverse flow injection analysis coupled with a long path length liquid waveguide capillary cell. Talanta **117**, 456–462 (2013)
70. P.S. Ellis, A.M.H. Shabani, B.S. Gentle, I.D. McKelvie, Field measurement of nitrate in marine and estuarine waters with a flow analysis system utilizing on-line zinc reduction. Talanta **84**(1), 98–103 (2011)
71. A.D. Beaton et al., Lab-on-chip measurement of nitrate and nitrite for in situ analysis of natural waters. Environ. Sci. Technol. **46**(17), 9548–9556 (2012)
72. N. Amini, I. McKelvie, An enzymatic flow analysis method for the determination of phosphatidylcholine in sediment pore waters and extracts. Talanta **66**(2), 445–452 (2005)
73. B. Paczosa-Bator, L. Cabaj, M. Raś, B. Baś, R. Piech, Potentiometric sensor platform based on a carbon black modified electrodes. Int. J. Electrochem. Sci. **9**, 2816–2823 (2014)
74. E. Lindner, B.D. Pendley, A tutorial on the application of ion-selective electrode potentiometry: an analytical method with unique qualities, unexplored opportunities and potential pitfalls; Tutorial. Anal. Chim. Acta **762**, 1–13 (2013)
75. A. Stortini, L. Moretto, A. Mardegan, M. Ongaro, P. Ugo, Arrays of copper nanowire electrodes: Preparation, characterization and application as nitrate sensor. Sens. Actuators B Chem. **207**, 186–192 (2015)
76. L.T. Duarte, C. Jutten, S. Moussaoui, A Bayesian nonlinear source separation method for smart ion-selective electrode arrays. IEEE Sens. J. **9**(12), 1763–1771 (2009)
77. P. Ciosek, W. Wróblewski, Potentiometric electronic tongues for foodstuff and biosample recognition—An overview. Sensors **11**(5), 4688–4701 (2011)
78. T. Öznülüer, B. Özdurak, H.Ö. Doğan, Electrochemical reduction of nitrate on graphene modified copper electrodes in alkaline media. J. Electroanal. Chem. **699**, 1–5 (2013)
79. Z. Chang, Y. Zhu, L. Zhang, S. Du, Measurement experiment and mathematical model of nitrate ion selective electrode, in *2013 Third International Conference on Instrumentation, Measurement, Computer, Communication and Control (IMCCC)* (IEEE, 2013), pp. 48–52
80. M.A.M. Yunus, S. Ibrahim, W.A.H. Altowayti, G.P. San, S.C. Mukhopadhyay, Selective membrane for detecting nitrate based on planar electromagnetic sensors array, in *2015 10th* Asian-Control *Conference (ASCC)* (IEEE, 2015), pp. 1–6
81. M.A.M. Yunus, S.C. Mukhopadhyay, Novel planar electromagnetic sensors for detection of nitrates and contamination in natural water sources. IEEE Sens. J. **11**(6), 1440–1447 (2011)
82. A.S.M. Nor, M.A.M. Yunus, S.W. Nawawi, S. Ibrahim, Low-cost sensor array design optimization based on planar electromagnetic sensor design for detecting nitrate and sulphate,"

in *2013 Seventh International Conference* on *Sensing Technology (ICST)* (IEEE, 2013), pp. 693–698
83. M.A.M. Yunus, S. Mukhopadhyay, A. Punchihewa, Application of independent component analysis for estimating nitrate contamination in natural water sources using planar electromagnetic sensor, in *2011 Fifth International Conference on Sensing Technology (ICST)* (IEEE, 2011), pp. 538–543
84. M.A.M. Yunus, S.C. Mukhopadhyay, S. Ibrahim, Planar electromagnetic sensor based estimation of nitrate contamination in water sources using independent component analysis. IEEE Sens. J. **12**(6), 2024–2034 (2012)
85. A.S.M. Nor, M. Faramarzi, M.A.M. Yunus, S. Ibrahim, Nitrate and sulfate estimations in water sources using a planar electromagnetic sensor array and artificial neural network method. IEEE Sens. J. **15**(1), 497–504 (2015)
86. M.M. Yunus, S.C. Mukhopadhyay, M. Rahman, N. Zahidin, S. Ibrahim, The selection of novel planar electromagnetic sensors for the application of nitrate contamination detection, in *Smart Sensors for Real-Time Water Quality Monitoring* (Springer, Berlin, 2013), pp. 171–195
87. M.M. Yunus, S.C. Mukhopadhyay, A. Punchihewa, S. Ibrahim, The effect of temperature factor on the detection of nitrate based on planar electromagnetic sensor and independent component analysis, in *Smart Sensing Technology for Agriculture and Environmental Monitoring* (Springer, Berlin, 2012), pp. 103–118
88. X. Wang, Y. Wang, H. Leung, S.C. Mukhopadhyay, M. Tian, J. Zhou, Mechanism and experiment of planar electrode sensors in water pollutant measurement. IEEE Trans. Instrum. Meas. **64**(2), 516–523 (2015)
89. A.V. Mamishev, K. Sundara-Rajan, F. Yang, Y. Du, M. Zahn, Interdigital sensors and transducers. Proc. IEEE **92**(5), 808–845 (2004)
90. M.E.E. Alahi, L. Xie, S. Mukhopadhyay, L. Burkitt, A temperature compensated smart nitrate-sensor for agricultural industry. IEEE Trans. Ind. Electron. **64**(9), 7333–7341 (2017)
91. M.E.E. Alahi, X. Li, S.C. Mukhopadhyay, L. Burkitt, A temperature compensated smart nitrate-sensor for agricultural industry. IEEE Trans. Ind. Electron. (2017)
92. M.E.E. Alahi, S.C. Mukhopadhyay, L. Burkitt, Imprinted polymer coated impedimetric nitrate sensor for real-time water quality monitoring. Sens. Actuators B Chem. **259**, 753–761 (2018)
93. M.E.E. Alahi, X. Li, S.C. Mukhopadhyay, L. Burkitt, Application of practical nitrate sensor based on electrochemical impedance spectroscopy, in *Sensors for Everyday Life* (Springer, Berlin, 2017), pp. 109–136
94. M.E.E. Alahi, A. Nag, S.C. Mukhopadhyay, L. Burkitt, A temperature-compensated graphene sensor for nitrate monitoring in real-time application. Sens. Actuators A **269**, 79–90 (2018)
95. M.E.E. Alahi, L. Xie, A.I. Zia, S.C. Mukhopadhyay, L. Burkitt, Practical nitrate sensor based on electrochemical impedance measurement, in *2016 IEEE International Instrumentation and Measurement Technology Conference Proceedings (I2MTC)* (IEEE, 2016), pp. 1–6
96. M.E.E. Alahi, N. Afsarimanesh, S.C. Mukhopadhyay, L. Burkitt, Development of the selectivity of nitrate sensors based on ion imprinted polymerization technique, in *2017 Eleventh International Conference on Sensing Technology (ICST)* (IEEE, 2017), pp. 1–6
97. M.E.E. Alahi, N. Pereira-Ishak, S.C. Mukhopadhyay, L. Burkitt, An internet-of-things enabled smart sensing system for nitrate monitoring. IEEE Internet Things J. (2018)
98. Y. Moo, M. Matjafri, H. Lim, C. Tan, New development of optical fibre sensor for determination of nitrate and nitrite in water. Optik-Int. J. Light Electron Optics **127**(3), 1312–1319 (2016)
99. A. Azmi, A.A. Azman, S. Ibrahim, M.A.M. Yunus, Techniques in advancing the capabilities of various nitrate detection methods: a review. Int. J. Smart Sens. Intell. Syst. **10**(2) (2017)
100. N. Amini, M. Shamsipur, M.B. Gholivand, K. Naderi, Electrocatalytic and new electrochemical properties of chloropromazine into silicaNPs/chloropromazine/Nafion nanocomposite: application to nitrite detection at low potential. Microchem. J. **131**, 43–50 (2017)
101. B. Mahieuxe, M. Carré, M. Viriot, J. André, M. Donner, Fiber-optic fluorescing sensors for nitrate and nitrite detection. J. Fluoresc. **4**(1), 7–10 (1994)

102. J. Camas-Anzueto, A. Aguilar-Castillejos, J. Castañón-González, M. Lujpan-Hidalgo, H.H. de León, R.M. Grajales, Fiber sensor based on Lophine sensitive layer for nitrate detection in drinking water. Opt. Lasers Eng. **60**, 38–43 (2014)
103. M.Y. Chong, M.Z.M. Jafri, L.H. San, T.C. Ho, Detection of nitrate ions in water by optical fiber, in *2012 International Conference on Computer and Communication Engineering (ICCCE)* (IEEE, 2012), pp. 271–273
104. K.S. Johnson, L.J. Coletti, H.W. Jannasch, C.M. Sakamoto, D.D. Swift, S.C. Riser, Long-term nitrate measurements in the ocean using the in situ ultraviolet spectrophotometer: sensor integration into the Apex profiling float. J. Atmos. Oceanic Technol. **30**(8), 1854–1866 (2013)
105. A. Lalasangi et al., Fiber Bragg grating sensor for detection of nitrate concentration in water. Sens. Transducers **125**(2), 187 (2011)
106. C. Munkholm, D.R. Walt, F.P. Milanovich, A fiber-optic sensor for CO_2 measurement. Talanta **35**(2), 109–112 (1988)
107. Y. Zhu, A. Wang, Miniature fiber-optic pressure sensor. IEEE Photonics Technol. Lett. **17**(2), 447–449 (2005)
108. S. Zhang, H. Chen, H. Fu, Fiber-optic temperature sensor using an optoelectronic oscillator, in *14th International Conference on Optical Communications and Networks (ICOCN)* (IEEE, 2015), pp. 1–3
109. F. Delport et al., Real-time monitoring of DNA hybridization and melting processes using a fiber optic sensor. Nanotechnology **23**(6), 065503 (2012)
110. P. Bhatia, B.D. Gupta, Fabrication and characterization of a surface plasmon resonance based fiber optic urea sensor for biomedical applications. Sens. Actuators B Chem. **161**(1), 434–438 (2012)
111. P. Bagade, A. Banerjee, S.K. Gupta, Evidence-based development approach for safe, sustainable and secure mobile medical app, in *Wearable Electronics Sensors* (Springer, Berlin, 2015), pp. 135–174
112. Flexible smart sensors and the future of health. https://www.engadget.com/2015/09/21/flexible-smart-sensors-and-the-future-of-health/
113. The state of flexible and printed electronics. http://www.printedelectronicsnow.com/issues/2015-03-01/view_features/the-state-of-flexible-and-printed-electronics
114. 2016–2026: market forecasts. http://www.idtechex.com/research/reports/wearable-sensors-2016-2026-market-forecasts-technologies-players-000470.asp
115. The wearable technology ecosystem: 2016–2030—opportunities, challenges, strategies, industry verticals and forecasts. http://www.openpr.com/news/348933/The-Wearable-Technology-Ecosystem-2016-2030-Opportunities-Challenges-Strategies-Industry-Verticals-And-Forecasts.html?__hstc = 197865264.638cea001f4f55aadbd731e528921f0a.1482980112036.1482980112036.1482980112036.1&__hssc=197865264.1.1482980112037&__hsfp=1381054282
116. Organic electronics will play a key role in increasing the utility of wearables. https://www.wearable-technologies.com/2016/03/organic-electronics-will-play-a-key-role-in-increasing-the-utility-of-wearables/
117. A. Nag, S. Mukhopadhyay, Smart home: recognition of activities of elderly for 24/7; coverage issues, in *Proceedings of the 2014 International Conference on Sensing Technology, Liverpool, UK*, vol. 2 (2014), pp. 480–489
118. V. Kafle, Y. Fukushima, H. Harai, Design and implementation of dynamic mobile sensor network platform. Commun. Mag. IEEE **53**(3), 48–57 (2015)
119. M.S. Khan, M.S. Islam, H. Deng, Design of a reconfigurable RFID sensing tag as a generic sensing platform toward the future internet of things. Internet Things J. IEEE **1**(4), 300–310 (2014)
120. M.E.E. Alahi, A. Nag, N. Afsari Manesh, S.C. Mukhopadhyay, J.K. Roy, *A* simple embedded sensor: Excitation and interfacing (Smart Sensors, Measurement and Instrumentation) (2017)
121. M.E.E. Alahi, S.C. Mukhopadhyay, Detection methodologies for pathogen and toxins: a review. Sensors (Switzerland) (2017). https://doi.org/10.3390/s17081885

122. H. Ghayvat, J. Liu, M. Alahi, S. Mukhopadhyay, X. Gui, Internet of things for smart homes and buildings: opportunities and challenges. Austr. J. Telecommun. Digital Econ. **3**(4), 33–47 (2015)
123. A. Ouadjaout et al., DZ50: energy-efficient wireless sensor mote platform for low data rate applications. Proc. Comput. Sci. **37**, 189–195 (2014)
124. A. Gluhak, S. Krco, M. Nati, D. Pfisterer, N. Mitton, T. Razafindralambo, A survey on facilities for experimental internet of things research. Commun. Mag. IEEE **49**(11), 58–67 (2011)
125. P. Mell, T. Grance, The NIST definition of cloud computing (2011)
126. A. Flammini, E. Sisinni, Wireless sensor networking in the internet of things and cloud computing era. Proc. Eng. **87**, 672–679 (2014)
127. I.F. Akyildiz, W. Su, Y. Sankarasubramaniam, E. Cayirci, Wireless sensor networks: a survey. Comput. Netw. **38**(4), 393–422 (2002)
128. E. Gaura, L. Girod, J. Brusey, M. Allen, G. Challen, *Wireless Sensor Networks: Deployments and Design Frameworks* (Springer Science & Business Media, 2010)
129. J. Chen, M. Díaz, L. Llopis, B. Rubio, J.M. Troya, A survey on quality of service support in wireless sensor and actor networks: requirements and challenges in the context of critical infrastructure protection. J. Netw. Comput. Appl. **34**(4), 1225–1239 (2011)
130. R. Moss Kanter, S.S. Litow, Informed and interconnected: A manifesto for smarter cities (2009)
131. S. Dirks, M. Keeling, *A Vision of Smarter Cities: How Cities Can Lead the Way into a Prosperous and Sustainable Future*, vol. 8 (IBM Institute for Business Value, 2009)
132. J.M. Shapiro, Smart cities: quality of life, productivity, and the growth effects of human capital. The review of economics and statistics **88**(2), 324–335 (2006)
133. G. Werner-Allen, P. Swieskowski, M. Welsh, Motelab: a wireless sensor network testbed, in *Proceedings of the 4th International Symposium on Information Processing in Sensor Networks* (IEEE Press, 2005), p. 68
134. M. Doddavenkatappa, M.C. Chan, A.L. Ananda, Indriya: A low-cost, 3D wireless sensor network testbed, in *Testbeds and Research Infrastructure. Development of Networks and Communities* (Springer, 2012), pp. 302–316
135. A. Burns et al., SHIMMER™—a wireless sensor platform for noninvasive biomedical research. Sens. J. **10**(9), 1527–1534 (2010)
136. S. Pirbhulal et al., A novel secure IoT-based smart home automation system using a wireless sensor network. Sensors (Switzerland) (2017). https://doi.org/10.3390/s17010069
137. G. Hackmann, W. Guo, G. Yan, Z. Sun, C. Lu, S. Dyke, Cyber-physical codesign of distributed structural health monitoring with wireless sensor networks. IEEE Trans. Parallel Distributed Syst. **25**(1), 63–72 (2014)
138. P. Sandeep et al., A novel secure IoT-based smart home automation system using a wireless sensor network. Sensors (2017). https://doi.org/10.3390/s17030606
139. Z. Riaz, M. Arslan, A.K. Kiani, S. Azhar, CoSMoS: a BIM and wireless sensor based integrated solution for worker safety in confined spaces. Autom. Construct. **45**, 96–106 (2014)
140. E. Cañete, J. Chen, M. Díaz, L. Llopis, B. Rubio, Sensor4PRI: a sensor platform for the protection of railway infrastructures. Sensors **15**(3), 4996–5019 (2015)
141. N. Poursafar, M.E.E. Alahi, S.C. Mukhopadhyay, Long-range wireless technologies for IoT applications: a review, in *11th International Conference on Sensing Technology (ICST)* (2017), pp. 310–315. ISBN 978-1-5090-6526-4
142. N. Afsarimanesh, M.E.E. Alahi, S.C. Mukhopadhyay, M. Kruger, A novel electrochemical biosensor for bone turnover detection based on molecular imprinting technology, in *11th International Conference on Sensing Technology (ICST)* (2017), pp. 6–11. ISBN 978-1-5090-6526-4
143. M.E.E. Alahi, N. Afsarimanesh, S.C. Mukhopadhyay, L. Burkitt, Development of the selectivity of nitrate sensors based on ion imprinted polymerization technique, in *2017 11th International Conference on Sensing Technology (ICST)* (2017), pp. 531–536. ISBN 978-1-5090-6526-4

144. N. Afsarimanesh, M. E. E. Alahi, S.C. Mukhopadhyay, M. Kruger, P.-L. Yu, Development of molecular imprinted polymer interdigital sensor for C-terminal telopeptide of type I collagen, in *2016 10th International Conference on Sensing Technology (ICST)* (IEEE, 2016), pp. 1–5
145. A. Nag, M.E.E. Alahi, S.C. Mukhopadhyay, IoT-based sensing system for phosphate detection using Graphite/PDMS sensors. Sens. Actuators A Phys. (2018)
146. N. Afsarimanesh, M.E.E. Alahi, S.C. Mukhopadhyay, M. Kruger. Development of IoT-based impedometric biosensor for point-of-care monitoring of bone loss. IEEE J. Emerg. Select. Topics Circ. Syst. (2018)
147. C. CC2420. MEMSIC Inc. https://inst.eecs.berkeley.edu/~cs150/Documents/CC2420.pdf
148. J. King, R. Bose, H.-I. Yang, S. Pickles, A. Helal, Atlas: a service-oriented sensor platform: Hardware and middleware to enable programmable pervasive spaces," in *31st IEEE Conference on Local Computer Networks* (IEEE, 2006), pp. 630–638
149. B. Zhou, S. Yang, T.H. Nguyen, T. Sun, K.T. Grattan, Wireless sensor network platform for intrinsic optical fiber pH sensors. Sens. J. **14**(4), 1313–1320 (2014)
150. S. Ferdoush, X. Li, Wireless sensor network system design using Raspberry Pi and Arduino for environmental monitoring applications. Proc. Comput. Sci. **34**, 103–110 (2014)

Chapter 3
Interdigitated Senor and Electrochemical Impedance Spectroscopy (EIS)

3.1 Introduction

This objective of this chapter is to explain the operating principle of the sensors in terms of their electrical behaviour. It also displays the phenomenon of impedance spectroscopy used to characterise the sensor and material under test (MUT). Electrochemical impedance spectroscopy (EIS) was used as the measurement tool for the developed prototypes to determine the nitrate in water. It also provides a details insight into EIS method. The response of the systems was determined as a function of frequency to calculate the changes occurring in linear and non-linear systems. EIS served as an excellent tool for measurement, as the sensors were operated at dynamic interfaces where specific system parameters were monitored. All the subsequent chapters and the related experiments followed the EIS method to characterise the sensor.

3.2 Planar Interdigital Sensors

Interdigital sensors are made of the comb-like or finger-like periodic pattern of parallel electrodes on a solid phase substrate. These electrodes are used to build up the capacitance which is related to the electric fields that penetrate the test material and carry useful information about the properties of the material sample [1]. One of the most significant benefits of the planar interdigital sensors is the single-side access to the Material Under Test (MUT). This property helps to penetrate the sample with magnetic, electric, or acoustic fields from only one side. The strength of the output signal can be controlled by changing the area of the sensor, the number of fingers, and the spacing between them. The capability of being used for non-destructive testing is another advantage of these sensors, which makes them more useful for inline testing and process control applications [2].

Md E. E. Alahi and S. C. Mukhopadhyay, *Smart Nitrate Sensor*, Smart Sensors, Measurement and Instrumentation 35, https://doi.org/10.1007/978-3-030-20095-4_3

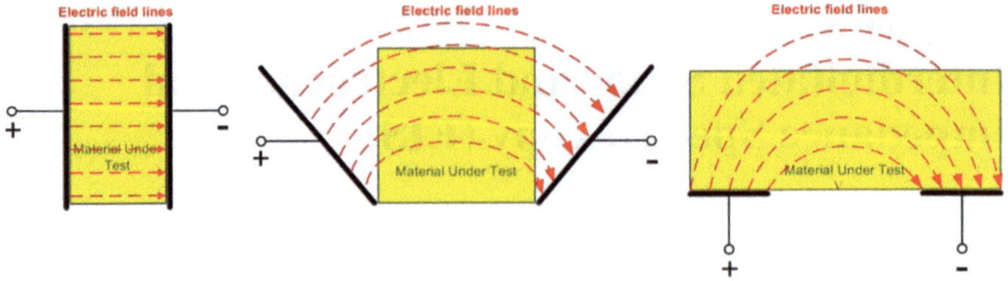

Fig. 3.1 Gradual transition from the parallel-plate capacitor to a planar capacitor

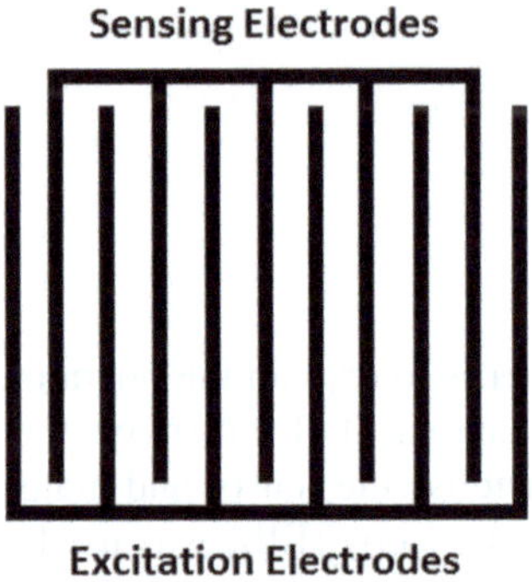

Fig. 3.2 Geometric structure of conventional planar interdigital sensor

Planar interdigital sensors follow the operating principle of parallel plate capacitors. Figure 3.1 shows a gradual transition from the parallel-plate capacitor to a planar, fringing field capacitor, where electrodes open up to provide single-side access to the MUT. The electrode pattern of the interdigital sensor can be repeated several times, to get a stronger signal and keep the signal-to-noise ratio in an acceptable range [3]. The configuration of the conventional interdigital sensor is shown in Fig. 3.2.

When an AC signal is applied as an excitation voltage to the terminals, an electric field is set up from positive to negative terminal. This electric field bulges through the test sample via excitation electrode and is received by the sensing electrode, which carries useful information about the properties of MUT such as impedance, density, chemical composition and so on. A change in the dielectric properties of materials is a function of chemical and physical properties of MUT. Figure 3.3 shows the electric field formed between positive and negative electrodes for different pitch lengths- the distance between two consecutive electrodes of the same polarity. As it is illustrated in the Fig. 3.3, different pitch length (l_1, l_2 and l_3) shows different penetration depth. The penetration depth improves by increasing the pitch length, at the cost of electric field strength that will get weaker with greater distance between the electrodes at the same potential (pitch length).

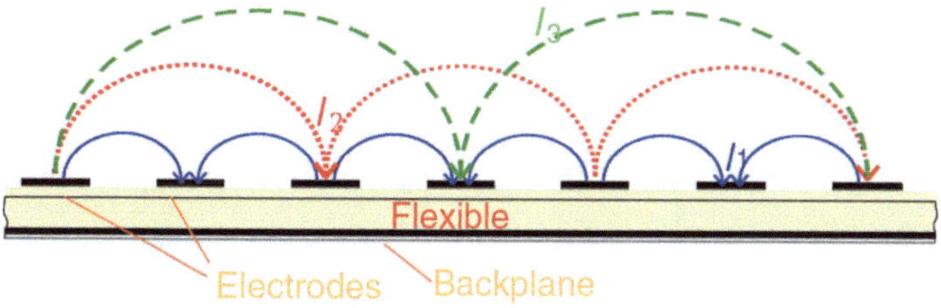

Fig. 3.3 Electric field formed for different pitch lengths [4]

3.3 Novel Planar Interdigital Sensors

Novel interdigital sensors are designed with more sensing electrodes compared to the excitation electrodes, to increase the penetration depth of the fringing electric field. Different geometries have been studied in research literature [5–7]. Figure 3.4 shows the excitation pattern for a multi-sensing electrode in interdigital sensor geometry.

The novel interdigital sensors have been fabricated based on different geometric parameters. Table 3.1 shows geometric parameters of four different interdigital sensors and Fig. 3.5 shows the schematic of 1-5-25 and 1-11-25 configuration of newly designed planer interdigital sensors [8].

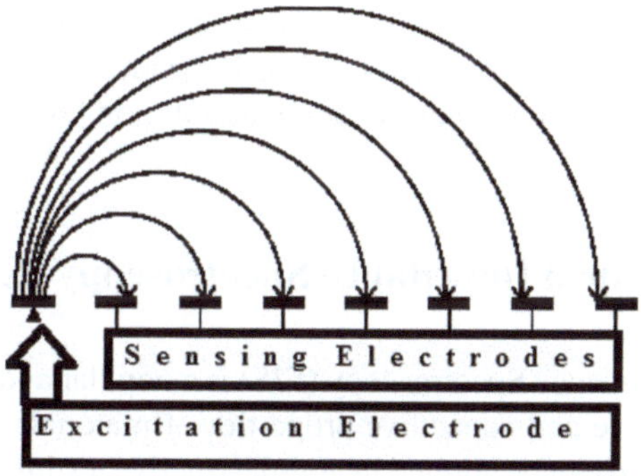

Fig. 3.4 Schematic excitation patterns for multi-sensing electrode interdigital sensors

Table 3.1 Geometric design parameters for four types of interdigital sensors

Sensor type	Pitch length (μm)	Number of sensing electrodes	Number of excitation electrodes	Sensing area (mm^2)
1-5-25	25	40	9	6.25
1-5-50	50	30	7	6.25
1-11-25	25	44	5	6.25
1-11-50	50	33	4	6.25

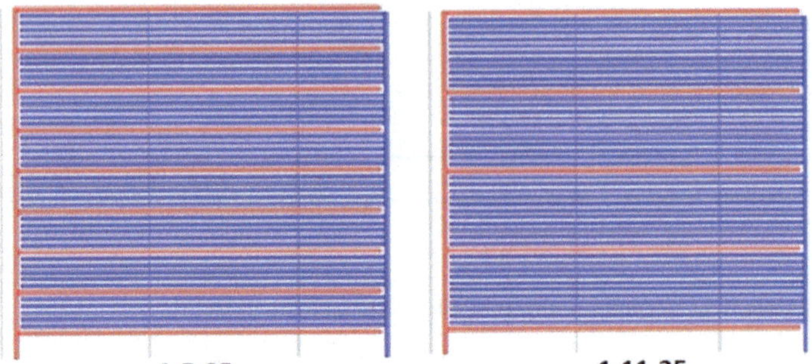

Fig. 3.5 1-5-25 and 1-11-25 configuration of Novel interdigital sensors

A sinusoidal signal is applied to the excitation electrodes of the proposed interdigital sensor. The electrical excitation or switching electric field bulges through the test-sample or the material under test via excitation electrode. The signal is received by the sensing electrode which carries useful information about the properties of the material under test [1, 2]. The material under test should be the close proximity of the sensing area to measure the electrical impedance clearly. These sensors have several applications in manufacturing process [9], environmental monitoring [7, 10–13], humidity and moisture sensing system [14, 15], photosensitive detection [16] and gas sensor [17]. A sensing system was developed based on the interdigital sensor to detect the dangerous contaminated chemical in seafood [5, 13]. Another one was developed based on the electrochemical impedance spectroscopy technique to monitor the presence of phthalates in aqueous solution [10–12].

3.4 Electrochemical Impedance Spectroscopy (EIS)

Electrochemical Impedance Spectroscopy (EIS) is a popular and powerful technique to measure the resistive and capacitive properties of materials by applying a small AC signal. EIS method can be applied in non-destructive testing, label-free detection and single-sided access for different chemical analysis. Different application of EIS had been reported such as the detection of fat contents in meat [18], biotoxins in shellfish [5], bacterial endotoxin in food [19], phthalates in water and juices [11], determination of the corrosive behavior of materials [20, 21] and analysis of electrical properties for soymilk coagulation process [22]. EIS has several applications in different research areas such as corrosion mechanisms [23], coating evaluation [24, 25], optimisation of batteries [26] and biosensing [27].

Among some methods available for impedance measurements, Frequency Response Analysis (FRA) has become a de facto standard for EIS measurement which is a rapid approach to evaluating the impedance variation in real-time. This technique measures the impedance of the system over a wide frequency range and

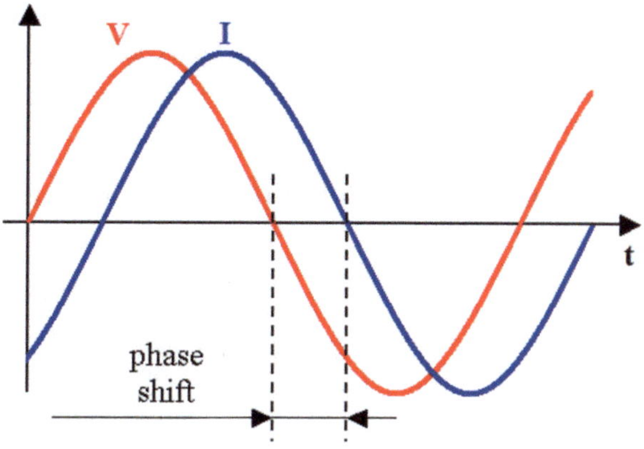

Fig. 3.6 Phase shift in current signal concerning the applied voltage

compares the results with reference data. In order to ensure system stability, linearity, and repeatability, this method is rendered viable only for a stable and reversible system in equilibrium [28–30].

3.4.1 *Basic Principles of EIS*

Electrochemical impedance can be measured by applying a small AC signal and then measuring the phase shift in the current signal concerning the applied potential. Electrochemical impedance is measured using a low excitation signal so that the cell's response is pseudo-linear. In a linear system, this current response to a sinusoidal excitation potential will result in a sinusoidal current at the same frequency but shifted in phase as shown in Fig. 3.6.

Impedance is defined as the measurement of the ability of a circuit to opposite the flow of electrical current when a voltage is applied. In an AC circuit, impedance is represented as a complex value, which involves real part (resistance) and imaginary part (reactance).

According to the fundamental Ohm's law in Eq. (3.1), impedance is defined as the ratio of potential, V and current, I:

$$R = \frac{V}{I} \tag{3.1}$$

The excitation signal can be explained as a function of time:

$$V_t = V_0 \sin \omega t \tag{3.2}$$

where V_t is the voltage difference at time t, V_0 is the amplitude of the signal, and the angular frequency is ω. It is given by ($\omega = 2\pi f$) expressed in radians/second and frequency, *f*, in hertz [31]. For a linear system, the response signal, I_t, has a phase shift, θ, with the amplitude of I_0 which can be explained by:

$$I_t = I_0 \sin(\omega t - \theta) \tag{3.3}$$

Equation (3.1) (Ohm's Law) can be used to calculate the impedance of the system given by:

$$\begin{aligned} Z = \frac{V_t}{I_t} &= \frac{V_0 \sin \omega t}{I_0 \sin(\omega t - \theta)} \\ &= Z_0 \frac{\sin(\omega t)}{\sin(\omega t - \theta)} \end{aligned} \tag{3.4}$$

The impedance, Z, now can be expressed in term of a magnitude of Z_0 and a phase shift, θ.

From Eq. (3.4), it can also be expressed in term of Euler's relationship given by:

$$e^{j\theta} = \cos\theta + \mathrm{j}\sin\theta \tag{3.5}$$

The impedance, Z, can be expressed in terms of potential, V, and current response, I, given by:

$$V_t = V_0 e^{j\omega t} \tag{3.6}$$

$$I_t = I_0 e^{j(\omega t - \theta)} \tag{3.7}$$

Therefore the impedance, Z:

$$Z(\omega) = \frac{V_t}{I_t} = \frac{V_0 e^{j\omega t}}{I_0 e^{j(\omega t - \theta)}} = Z_0 e^{j\theta} \tag{3.8}$$

$$= Z_0(\cos\theta + j\sin\theta) \tag{3.9}$$

The impedance now is in the form of the real part ($Z_0 \cos\theta$) and imaginary part (Z_0 $\sin\theta$). The Calculated impedance characteristics can be analysed using the Nyquist plot and Bode plot.

3.4.2 Data Representation in Nyquist Plot and Bode Plot

The Nyquist plot is an accessible format for evaluating electrochemical impedance data such as electrolytic solution resistance (R_s), electrode polarisation resistance (R_p) and double layer capacitance (C_{dl}). These parameters will be discussed in the following sections. Nyquist plot represents an imaginary impedance component against

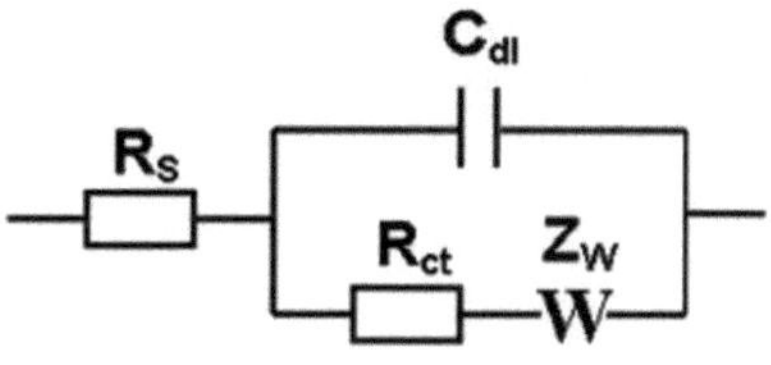

Fig. 3.7 Randle's equivalent circuit model

real impedance component at each excitation frequency. The Nyquist plot offers several advantages. The effects of solution resistance can be observed easily using this format. One major drawback of the Nyquist plot is that there is no information about the frequency in the Nyquist plot which makes it difficult to calculate the double layer capacitance [29].

The Bode plot represents the absolute impedance and the phase shift in the frequency domain. Since frequency appears at one of the axes, the effect of frequency on the impedance and phase shift can be studied. R_s, R_p, C_{dl} and frequency values, with phase shift can be studied where they are maximum/minimum. Bode plot is desirable to provide more explicit description of electrochemical cell's which shows frequency dependent behaviour compared to Nyquist plot.

3.4.3 Randle's Electrochemical Cell Equivalent Circuit Model

Randle's equivalent circuit is the most frequently discussed equivalent circuit [32] used to interpret EIS experimental results in electrical form as shown in Fig. 3.7. It consists of solution resistance R_s in series connected to a parallel combination of double layer capacitance C_{dl} to the charge transfer resistance R_{ct} in series with Warburg impedance Z_w [27].

This model was introduced by Randle in 1947 [33], and it can be used to describe both kinetics and diffusion processes taking place at the electrode-electrolyte interface.

The Nyquist plot from the equivalent circuit consists of a semi-circular region followed by a 45° straight line, as shown in Fig. 3.8. In this model, the impedance of a faradaic reaction consists of an active charge transfer resistance R_{ct}, and a specific electrochemical element of diffusion, which is called the Warburg element. The semi-circular region represents a slower charge transfer at higher frequencies whereas the straight line describes a faster mass-transfer at lower frequencies. R_{ct} can be calculated by extrapolating the semicircle to Z_{real} axis. Solution resistance R_s can be calculated by reading the real axis value at the high-frequency intercept, which is the intercept near the origin of the Nyquist plot. R_{ct} can be calculated by extrapolating the semicircle to Z_{real} axis as illustrated in Fig. 11. C_{dl} can be obtained from the frequency at the maximum of semicircle portion in the Nyquist plot using $\omega = 1/R_{ct}\ C_{dl}$ [32].

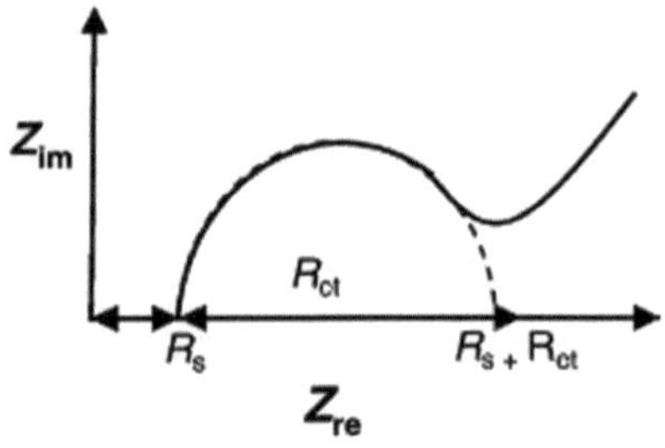

Fig. 3.8 The Nyquist plot for the Randle's equivalent circuit

The absolute impedance as a function of frequency is given as:

$$Z(\omega) = R_s + \frac{R_{ct}}{1 + \omega^2 R_{ct}^2 C_{dl}^2} - \frac{j\omega R_{ct}^2 C_{dl}}{1 + \omega^2 R_{ct}^2 C_{dl}^2} \tag{3.10}$$

where the real part (Z′) is given by:

$$Z'(\omega) = R_s + \frac{R_{ct}}{1 + \omega^2 R_{ct}^2 C_{dl}^2} \tag{3.11}$$

moreover, the imaginary part (Z″) is given by:

$$Z''(\omega) = -\frac{\omega R_{ct}^2 C_{dl}}{1 + \omega^2 R_{ct}^2 C_{dl}^2} \tag{3.12}$$

The rate of an electrochemical reaction can be strongly influenced by the diffusion of reactants towards, or away from the electrode-electrolyte interface. This situation can exist when the electrode is covered with adsorbed solution components or a selective coating. An additional element called Warburg impedance, Z_W, appears in series with resistance R_{ct}. Mathematically Warburg impedance is given by:

$$Z_W = \frac{\sigma_w}{\sqrt{j\omega}} \tag{3.13}$$

where, σ_w is the Warburg diffusion coefficient. The diffusion of reactants to the electrode surface is a slow-paced process, which can happen at low frequencies only. However, at higher frequencies, the reactants do not have enough time to diffuse. The slope of this line gives the Warburg diffusion coefficient.

3.5 Chapter Summary

This chapter gives a summary of the working principle of the sensor prototypes. Electrochemical Impedance Spectroscopy was used as the technique to monitor the responses of the sensor for nitrate measurement. The impedance analysers were chosen to do the measurements for the linear and non-linear systems in the succeeding chapters.

References

1. A.S. Abu-Abed, R.G. Lindquist, Capacitive interdigital sensor with inhomogeneous nematic liquid crystal film. Prog. Electromagn. Res. B **7**, 75–87 (2008)
2. A.V. Mamishev, K. Sundara-Rajan, F. Yang, Y. Du, M. Zahn, Interdigital sensors and transducers. Proc. IEEE **92**(5), 808–845 (2004)
3. Y. Chen, C. Zhu, M. Cao, T. Wang, Photoresponse of S_nO_2 nanobelts grown in situ on interdigital electrodes. Nanotechnology **18**(28), 285502 (2007)
4. M.S.A. Rahman, S.C. Mukhopadhyay, P.-L. Yu, Novel planar interdigital sensors, in *Novel Sensors for Food Inspection: Modelling, Fabrication and Experimentation* (Springer, 2014), pp. 11–35
5. A.M. Syaifudin, K. Jayasundera, S. Mukhopadhyay, A low cost novel sensing system for detection of dangerous marine biotoxins in seafood. Sens. Actuators B: Chem. **137**(1), 67–75 (2009)
6. S. Mukhopadhyay, P. Yu, Electromagnetic field computation using COMSOL Multiphysics to evaluate the performance of novel interdigital sensors, in *Applied Electromagnetics Conference (AEMC), 2009* (IEEE, 2009), pp. 1–4
7. M. Yunus, S. Mukhopadhyay, K. Jayasundera, A novel planar interdigital sensor for environmental monitoring, in *Sensors, 2009 IEEE* (IEEE, 2009), pp. 105–110
8. H. Arwin, Application of ellipsometry techniques to biological materials. Thin Solid Films **519**(9), 2589–2592 (2011)
9. Y. Yang et al., In situ on-line cure monitoring of composites by embedded interdigital sensor, in *16th European Conference on Composite Materials (ECCM-16)* (Universidad de Sevilla, School of Engineering, 2014)
10. A.I. Zia et al., Electrochemical impedance spectroscopy based MEMS sensors for phthalates detection in water and juices, J. Phys. Conf. Ser. **439**(1), 012026 (2013)
11. A.I. Zia, S.C. Mukhopadhyay, P.-L. Yu, I. Al-Bahadly, C.P. Gooneratne, J. Kosel, Rapid and molecular selective electrochemical sensing of phthalates in aqueous solution. Biosens. Bioelectron. **67**, 342–349 (2015)
12. A. Zia, S. Mukhopadhyay, I. Al-Bahadly, P. Yu, C.P. Gooneratne, J. Kosel, Introducing molecular selectivity in rapid impedimetric sensing of phthalates, in *Instrumentation and Measurement Technology Conference (I2MTC) Proceedings, 2014 IEEE International* (IEEE, 2014), pp. 838–843
13. A. Syaifudin, K. Jayasundera, S. Mukhopadhyay, A novel planar interdigital sensor based sensing and instrumentation for detection of dangerous contaminated chemical in seafood, in *Instrumentation and Measurement Technology Conference, 2009. I2MTC'09, IEEE* (IEEE, 2009), pp. 701–706
14. J. Fischer, H. Dejmkova, J. Barek, Electrochemistry of pesticides and its analytical applications. Curr. Org. Chem. **15**(17), 2923–2935 (2011)
15. M. Khafaji, S. Shahrokhian, M. Ghalkhani, Electrochemistry of levo-thyroxin on edge-plane pyrolytic graphite electrode: application to sensitive analytical determinations. Electroanalysis **23**(8), 1875–1880 (2011)
16. L. Li et al., In situ growth of ZnO nanowires on Zn comb-shaped interdigitating electrodes and their photosensitive and gas-sensing characteristics. Mater. Res. Bull. **47**(12), 3971–3975 (2012)
17. M. Dhull, A. Arora, Design of MEMS based microheater for enhanced efficiency of gas sensors. J. Therm. Eng. Appl. **2**(2), 16–21 (2015)

18. S.C. Mukhopadhyay, C.P. Gooneratne, A novel planar-type biosensor for noninvasive meat inspection. Sens. J., IEEE **7**(9), 1340–1346 (2007)
19. M.S.A. Rahman et al., Detection of bacterial endotoxin in food: New planar interdigital sensors based approach. J. Food Eng. **114**(3), 346–360 (2013)
20. C. Xhoffer, K. Van den Bergh, H. Dillen, Electrochemistry: a powerful analytical tool in steel research. Electrochim. Acta **49**(17), 2825–2831 (2004)
21. C.V. Vidal, A.I. Muñoz, Effect of physico-chemical properties of simulated body fluids on the electrochemical behaviour of CoCrMo alloy. Electrochim. Acta **56**(24), 8239–8248 (2011)
22. X. Li, K. Toyoda, I. Ihara, Coagulation process of soymilk characterized by electrical impedance spectroscopy. J. Food Eng. **105**(3), 563–568 (2011)
23. C. Liu, Q. Bi, A. Leyland, A. Matthews, An electrochemical impedance spectroscopy study of the corrosion behaviour of PVD coated steels in 0.5 N NaCl aqueous solution: part II.: EIS interpretation of corrosion behaviour. Corros. Sci. **45**(6), 1257–1273 (2003)
24. A. Ghasemi, V. Raja, C. Blawert, W. Dietzel, K. Kainer, Study of the structure and corrosion behavior of PEO coatings on AM50 magnesium alloy by electrochemical impedance spectroscopy. Surf. Coat. Technol. **202**(15), 3513–3518 (2008)
25. R.M. Souto, M.M. Laz, R.L. Reis, Degradation characteristics of hydroxyapatite coatings on orthopaedic TiAlV in simulated physiological media investigated by electrochemical impedance spectroscopy. Biomaterials **24**(23), 4213–4221 (2003)
26. D. Andre, M. Meiler, K. Steiner, C. Wimmer, T. Soczka-Guth, D. Sauer, Characterization of high-power lithium-ion batteries by electrochemical impedance spectroscopy. I. Experimental investigation. J. Power Sources **196**(12), 5334–5341 (2011)
27. F. Lisdat, D. Schäfer, The use of electrochemical impedance spectroscopy for biosensing. Anal. Bioanal. Chem. **391**(5), 1555–1567 (2008)
28. B.-Y. Chang, S.-M. Park, Electrochemical impedance spectroscopy. Annu. Rev. Anal. Chem. **3**, 207–229 (2010)
29. G. Instruments, Basics of electrochemical impedance spectroscopy. *G. Instruments, Complex impedance in Corrosion,* pp. 1–30 (2007)
30. I.I. Suni, Impedance methods for electrochemical sensors using nanomaterials. TrAC Trends Anal. Chem. **27**(7), 604–611 (2008)
31. A. Zia, Smart electrochemical sensing system for the real time detection of endocrine disrupting compounds and hormones, Thesis of Ph.D.
32. J.R. Macdonald, E. Barsoukov, Impedance spectroscopy: theory, experiment, and applications. History **1**, 8 (2005)
33. J.E.B. Randles, Kinetics of rapid electrode reactions. Discuss. Faraday Soc. **1**, 11–19 (1947)

Chapter 4
Temperature Compensation for Low Concentration Nitrate Measurement

4.1 Introduction

This chapter explains the design and development of a portable sensing system that could be used in situ to detect the concentrations of nitrate present in the groundwater. Electrochemical Impedance Spectroscopy (EIS) was employed for identifying and displaying nitrate concentrations by using a planar interdigital sensor which was immersed in the stream water samples. Calibration samples were prepared by serial dilution method from the stock nitrate solution in the laboratory and sample water for testing were collected from different sources of surface water. The developed system tests the collected sample waters and validated by standard equipment. The real part of impedances was taken to develop the standard calibration curve which showed an excellent linear relationship for different nitrate concentrations (range from 0.1 to 0.5 mg/L). The comparative results are promising which announced the extraordinary potential of the developed system for its in situ nitrate measurement with real-time monitoring. The temperature compensation capability is added to the system to compensate for any change of measurement due to temperature change in water. WiFi-based Internet of Things (IoT) has been incorporated with the developed system and providing its internet connectivity with the sensing system. The system is accomplished by sending data directly from the sampling location to an IoT-based cloud server. Therefore, the developed system will be suitable to develop real-time Wireless Sensor Network (WSN) systems in the future. The developed system has the potential to monitor the impact of the industrial, agricultural or urban activity on water quality, in real-time.

Md E. E. Alahi and S. C. Mukhopadhyay, *Smart Nitrate Sensor*, Smart Sensors, Measurement and Instrumentation 35, https://doi.org/10.1007/978-3-030-20095-4_4

4.2 Experimental Setup

4.2.1 MEMS-Based Sensor

An interdigital sensor with thin-film coated layer of Parylene was used as shown in Fig. 4.2. The purpose of using Parylene on the sensing surface is to protect the sensor from oxidation due to moisture and Faradic currents. The schematic diagram of the sensor used for the experiment is shown in Figs. 4.1 and 4.2. The configuration is 1-5-50, which means a repeated pattern of five sensing electrodes is present on one excitation electrode with a distance of 50 μm between two consecutive electrodes [1]. The working principle and EIS measurement of the sensor was explained in Chap. 3.

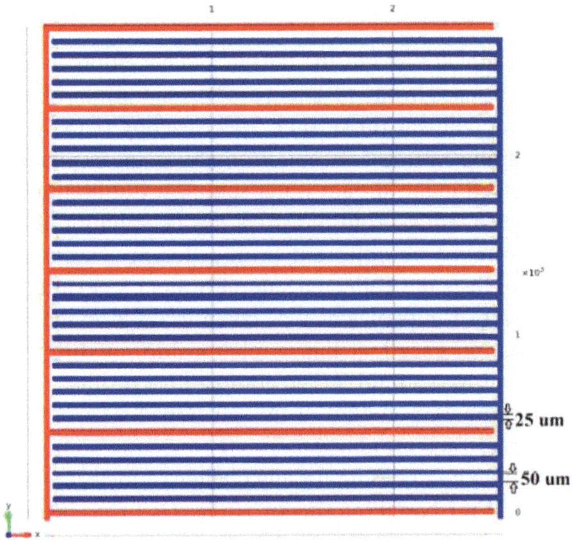

Fig. 4.1 Sensor configurations (1-5-50)

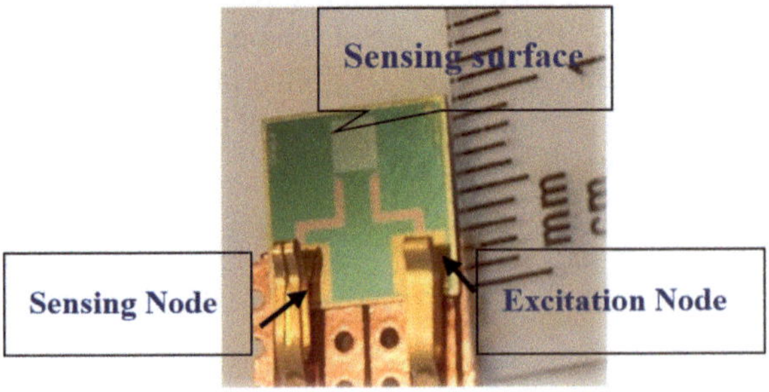

Fig. 4.2 Interdigital sensor with Parylene coating

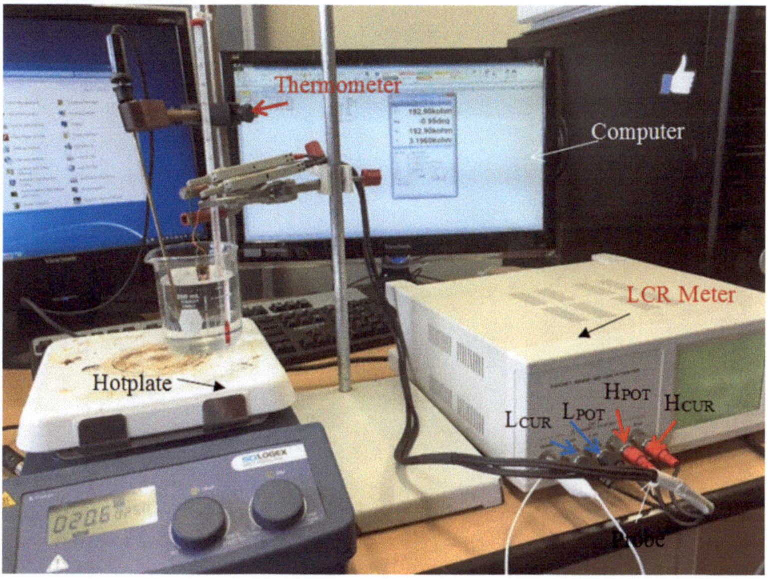

Fig. 4.3 Experimental arrangement for temperature measurement

4.2.2 Temperature Measurement

The mobility of ions in water can be varied due to the variation of temperature in water. Therefore, it is essential to measure the behaviour of the sensor at different temperatures. Figure 4.3 shows the experimental setup which involved a high precision Hioki 3522-50 LCR meter, Hioki 4-terminal probe 9140, mercury thermometer, SCILOGEX MS 7-H550 Digital Hotplate and computer for data acquisition. The mercury thermometer was dipped in deionised water (MilliQ® was obtained from MILLIPORE water system (USA)—18 MΩ cm) to obtain a continuous temperature reading. Meanwhile, the sensing surface was also dipped in water and collected data when the temperature reached a steady value. The frequency was swapped from 1 to 10,000 Hz, to characterize the sensor under different temperatures. The data was collected in an excel sheet for further analysis.

4.2.3 Nitrate Measurement

Figure 4.4 shows the experimental setup for NO_3-N concentration measurement. Experiments were performed in laboratory conditions in constant temperature and humidity. The 12-Volt DC battery was used as a power source for the main circuit board, water pump, and solenoid valve. Arduino Fio was used to generate the sinusoidal waveform that was the input signal for the sensor and performs data acquisi-

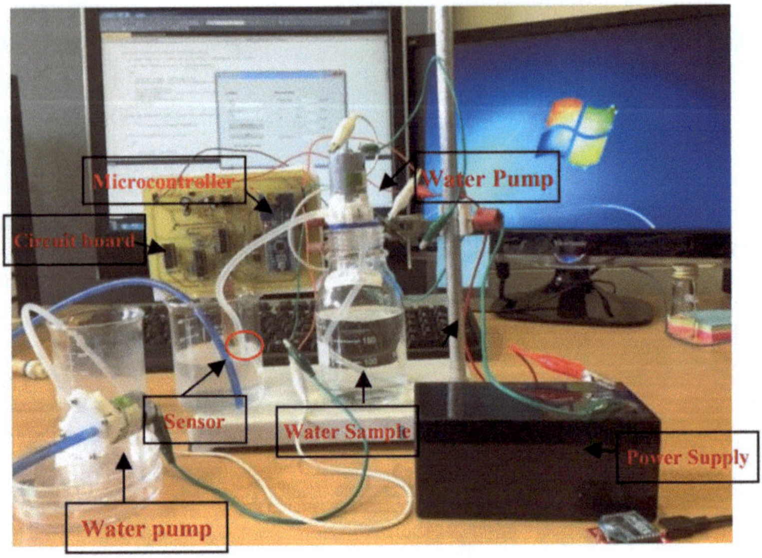

Fig. 4.4 Experiment setup for nitrate detection

tion. The sensing part of the sensor was immersed into calibration standard samples with different NO_3-N concentrations to create references and later water samples collected from a stream near Palmerston North; New Zealand was tested based on the calibration plot obtained from the standard solution. The electronic circuit was designed to get the input and output voltage of the sensor and to detect the phase shift between the incident and the signal received through the sensor. The ions in the nitrate get polarised towards the plates according to their charge [2]. The experiments were conducted in laboratory conditions; two water pumps were used. The pumps were connected to the motor from one being used to pump the water sample into the container for measurement at every 15 min, and another one was used for discharging the sampled water from the container after one minute. The microcontroller calculated the sensor's impedance. The detailed explanation is described in the succeeding section.

4.3 Sensing System

The sensing system was assembled, as a combination of different circuits performing required operations as described in the following sections.

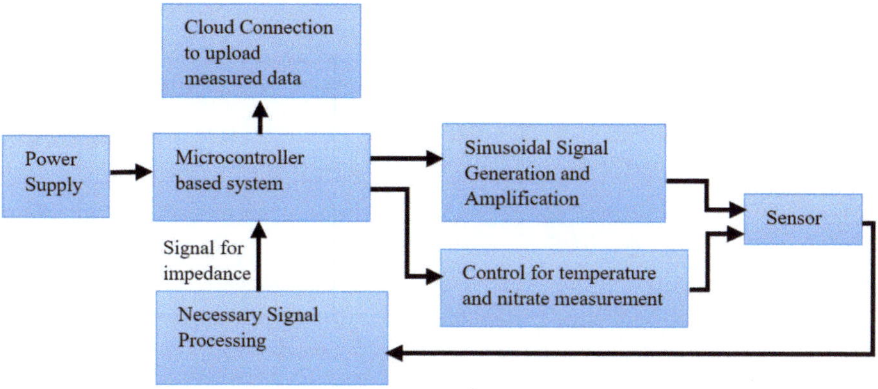

Fig. 4.5 Block diagram of the designed system

4.3.1 Block Diagram of the Sensing System

Figure 4.5 shows the block diagram of the designed system. In this system, the sinusoidal waveform was generated by using PWM (pulse width modulation) output combined with a bandpass filter, which is based on the concept of the Direct Digital Synthesis (DDS) method. This method was implemented by breaking a waveform into discrete points digitally [3]. Two hundred and fifty-six (8 bits) points were used to produce a sinusoidal waveform that gave a compromise between resolution and frequency. The operating frequency was fixed at 122.5 Hz.

4.3.2 The Signal Generator

The sinusoidal waveform was generated by using PWM output combined with a band-pass filter based on the concept of the Direct Digital Synthesis (DDS) method. DDS was implemented by breaking a waveform into discrete points digitally; the more points were taken, and the higher precision of DAC [4]. 256 (8 bits) points was used to produce a sinusoidal waveform that gave a compromise between resolution and frequency. As resolution increased, the maximum achievable frequency decreased. With the full resolution (255), the maximum PWM frequency was reached. The duty cycle and the pulse width of the digital signal were optimised [5] to generate different analogue levels. By reading these points to generate a sinusoidal waveform, the value of Output Compare Registers (OCR0) was changed.

The bandpass filter shown in Fig. 4.6 was employed, and the cut-off frequency was set between 106 and 338 Hz for generating the smooth sine wave. After filtering, the smooth sine wave was generated, and its frequency was fixed at 120 Hz defined by the value of Output Compare Registers (OCR1A as Timer 1 used).

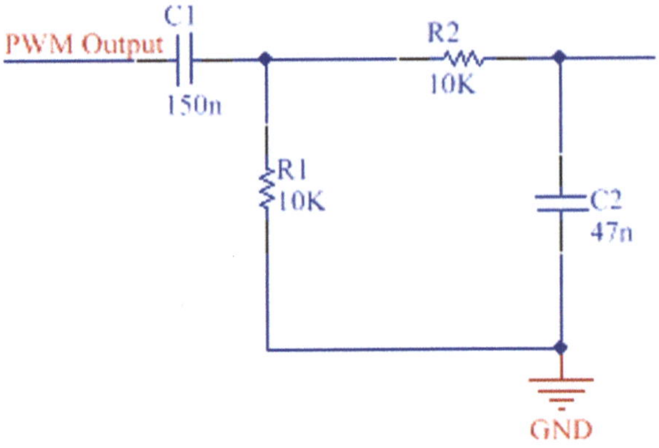

Fig. 4.6 Band-pass filter connected to PWM output

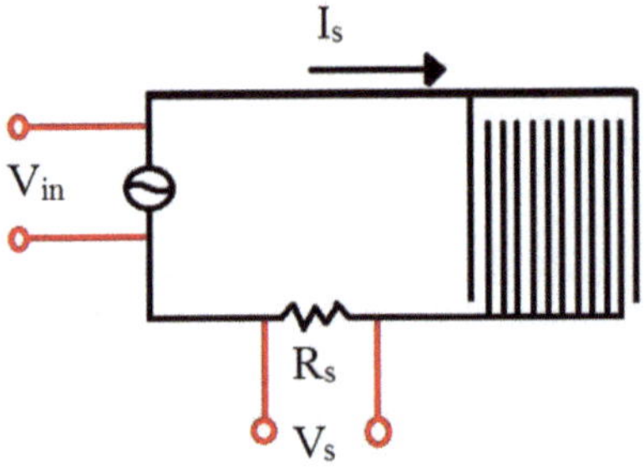

Fig. 4.7 Equivalent circuit diagram of interdigital sensor

4.3.3 Frequency Response Analysis Circuit

Due to the capacitive nature of the interdigital sensor, a resistor was connected to the sensor in series to determine the output voltage across the sensor. Figure 4.7 illustrates the equivalent circuit diagram of the interdigital sensor.

Therefore, the current (I_s) flow through the sensor can be calculated by:

$$I_s = \frac{V_{in}}{Z} \tag{4.1}$$

where V_{in} is the voltage applied to the sensor, Z is the total impedance of the circuit.

The sensing voltage V_s which measured across the series resistance can be calculated by:

$$V_s = I_s \times R_s = \frac{V_{in}}{Z} \times R_s \tag{4.2}$$

where, R_s is the series resistance, used for measurement of sensing voltage.

The active low-pass filter is shown in Fig. 4.8 was employed to reduce the noise and amplify the signal with gain as 10.

The impedance of the sensor can be calculated from the following relationship:

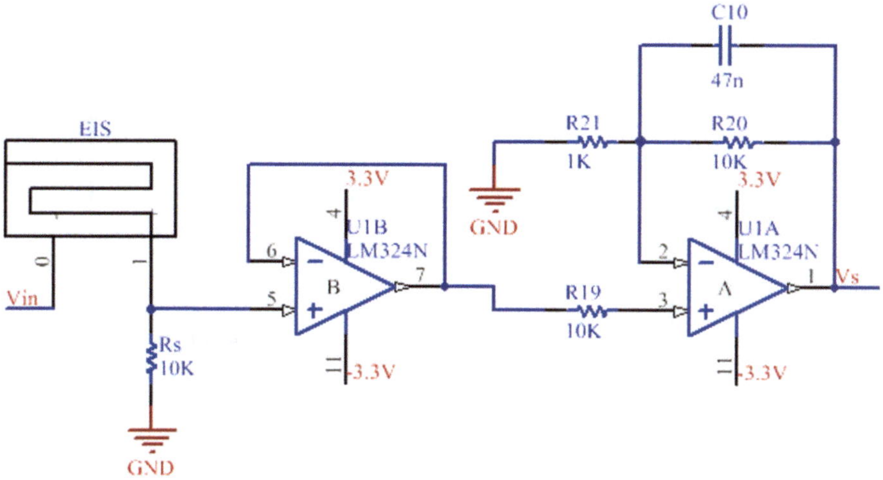

Fig. 4.8 An amplification circuit

$$Z = \frac{V_{in}}{I_s} = \frac{V_{in}}{V_s} \times R_s \tag{4.3}$$

The voltage applied to the sensor (V_{in}), and the voltages across the sensor (V_s) were measured, and microcontroller calculated Z. Series Resistor (R_s) was used at 10 KΩ, which is significantly small compared to Z. The precision rectifier is shown in Fig. 4.9 was employed to reduce the voltage loss because the V_{sensor} was tiny. The output of V_s was full sine wave fed into two operational amplifiers as shown in Fig. 4.10. Only positive half waves were generated from the first op-amp, as is seen in Fig. 4.9. The second op-amp worked as a subtractor, $V_s = V_{pos}/2$, when $V_s > 0$, $V2 = V_{pos} - V_s = V_{pos}/2$; when $V_s < 0$, $V_2 = 0-(-V_s) = V_{pos}$; Therefore the output V_2 is full wave rectified and fed into the microcontroller.

The R_s is a resistive element, so there is no phase difference between $V_{s.}$ And I_s. The phase angle between V_{in} and I_s is the same phase angle between V_{in} and V_s. LM339 consisting of four voltage comparators was used as hysteresis comparator to measure the phase angle (ϕ) between V_{in} and I_s with zero reference voltage as seen in Fig. 4.10. The purpose of using LM339 is to convert the V_{in} and V_s to square wave to calculate the phase angle in between them. It is also useful to reduce undesirable transitions caused by the noisy input signal.

The real part (R) and imaginary part (X) of impedance can be calculated by:

$$R = Z \times \cos \emptyset \tag{4.4}$$

$$X = Z \times \sin \emptyset \tag{4.5}$$

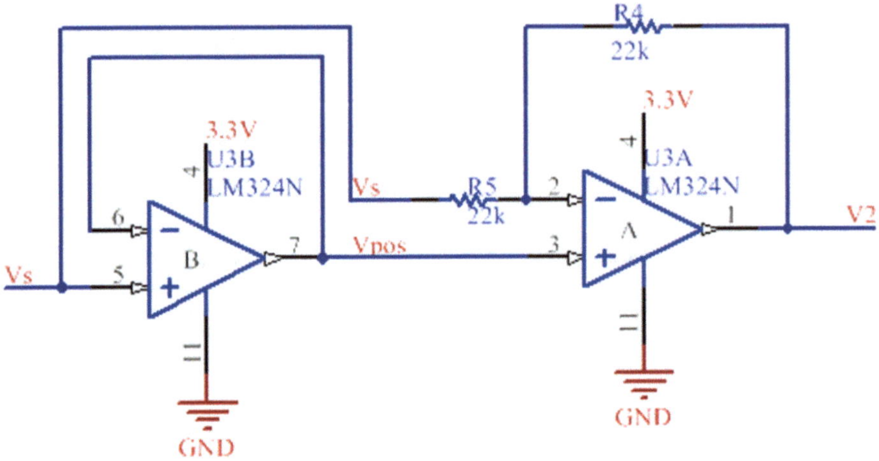

Fig. 4.9 Full rectifier circuit

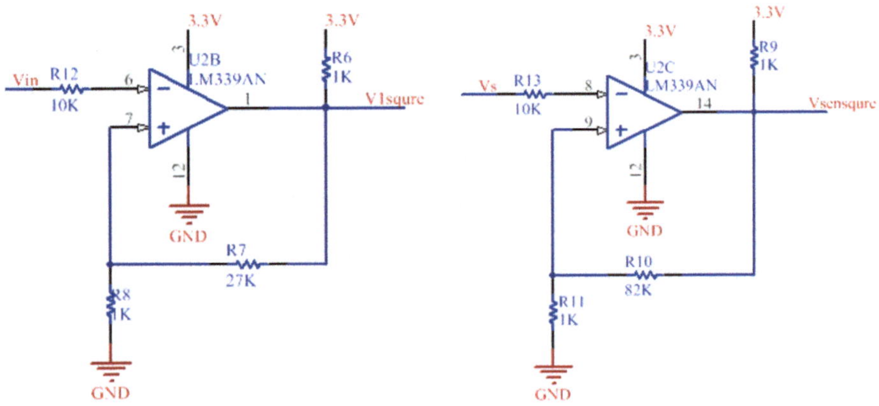

Fig. 4.10 Zero-crossing circuits

where R and X are the resistance and reactance of the circuit. Therefore, the resistive part of the sensor (R_{sensor}) is calculated by:

$$R = R - R_s \tag{4.6}$$

4.3.4 Controlling of Pump and Solenoid Valve

For field trial, it was planned to pump water into the sample container and to collect the impedance data. The water then was pumped out of the sample container. It was

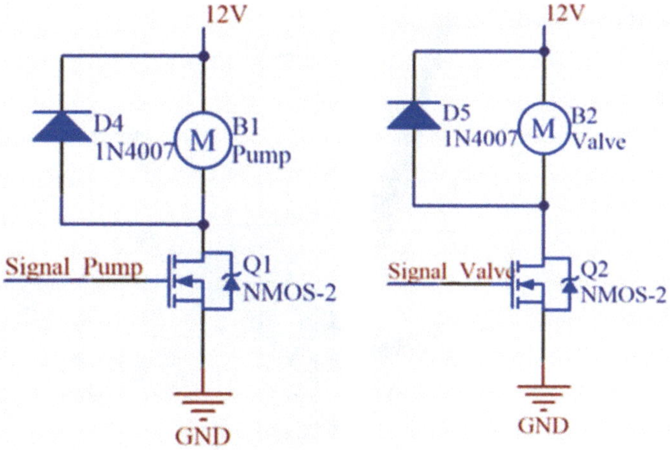

Fig. 4.11 Water pump and solenoid valve for sample flow control

done to avoid the sensor to be continuously dipped in water. The water pump and solenoid valve were switched by two MOSFETs which were controlled by microcontroller signals shown in Fig. 4.11 at particular time intervals. As the water pump required an input voltage of 5 volts, the PWM with 40% duty cycle was applied to the microcontroller output pin and then provided 5-volt excitation to the water pump required for its operation.

4.3.5 IoT-Based Smart System

IoT [6] offers promising solutions to transform the operation and role of existing industrial technologies [7–10]. IoT is already having an impact in the areas of agriculture, food processing, environmental monitoring, security surveillance [11–14] and others [15]. The proposed Arduino Yun has integrated WiFi which can provide instant connectivity to the Internet. WiFi offers high bandwidth, large coverage area, non-line-of-sight transmission, natural expansion, cost-effectiveness, robustness and small distribution of Links [16]. An external antenna (2.4 GHz) is added to increase the transmission signal strength. The collected data is transmitted to Thingspeak [17] which is the open data platform for the IoT. HTTP POST [18] protocol has been used to send data directly to the specified server. Figure 4.12 shows the final IoT-based smart sensing system with a smart sensor which has been used to measure nitrate and upload the data on the designated website.

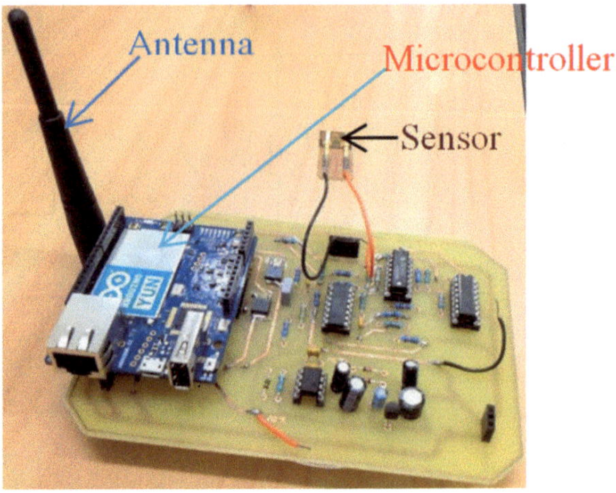

Fig. 4.12 IoT enabled smart sensing system

4.4 Results and Discussions

4.4.1 Measurement of Temperature

The same sensor was used to measure the temperature of water [19]. The real impedance (R_s) and imaginary impedance (X) were evaluated and plotted in Fig. 4.13. The X-axis shows the resistance part of impedance in ohms (Ω), and the Y-axis shows the reactance part of the impedance in ohms (Ω). The total impedance is decreased with the increase in temperature. The Nyquist plot (Fig. 4.13) indicates the impedance at a specific frequency.

Figure 4.14 shows the relationship between the imaginary part of impedance (X) and the frequency. It is seen from the figure that the resistance is decreased with increasing temperature, and the sensitive range is significant till 400 Hz. Figure 4.15 shows the real part of impedance (R_s) of deionised water to changing temperature, plotted against frequency. The dielectric properties of the deionised water at different temperatures was measured by using this approach. The experiment was repeated five times to observe impedance behaviour and to calculate average results.

Figure 4.16 shows the measured resistance as a function of the ambient temperature, which is almost linear. The slope can be calculated by Eq. (4.7):

$$Slope = \frac{\Delta R}{\Delta T} = \frac{R_{40} - R_{10}}{T_{40} - T_{10}} \tag{4.7}$$

where, R_{40}: The resistance value measured at 40 °C, R_{10}: The resistance value measured at 10 °C, T_{40}: The temperature at 40 °C, and T_{10}: The temperatures at 10 °C. Therefore, the slope for the resistance part is calculated as: −(42,392.83 − 118,037.76)/(40-10) = −2521.498. It is to be noted that all the resistance of

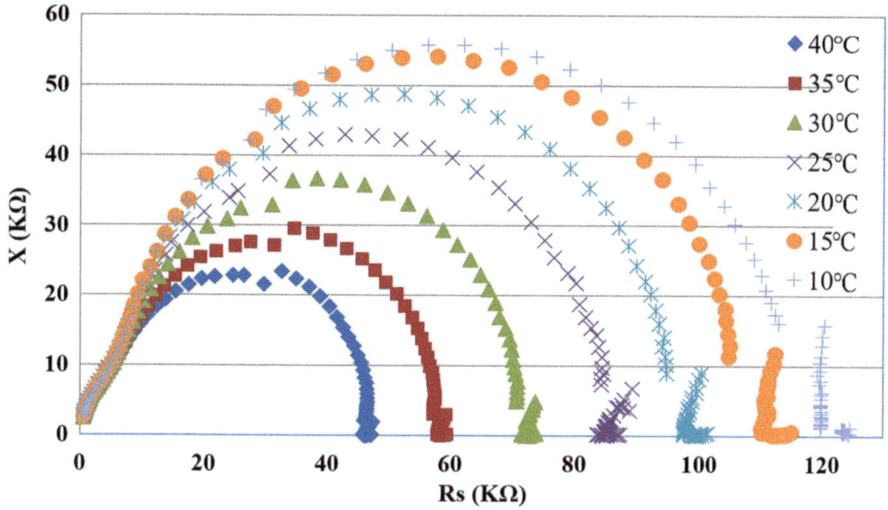

Fig. 4.13 Nyquist plot of impedance at variable temperatures

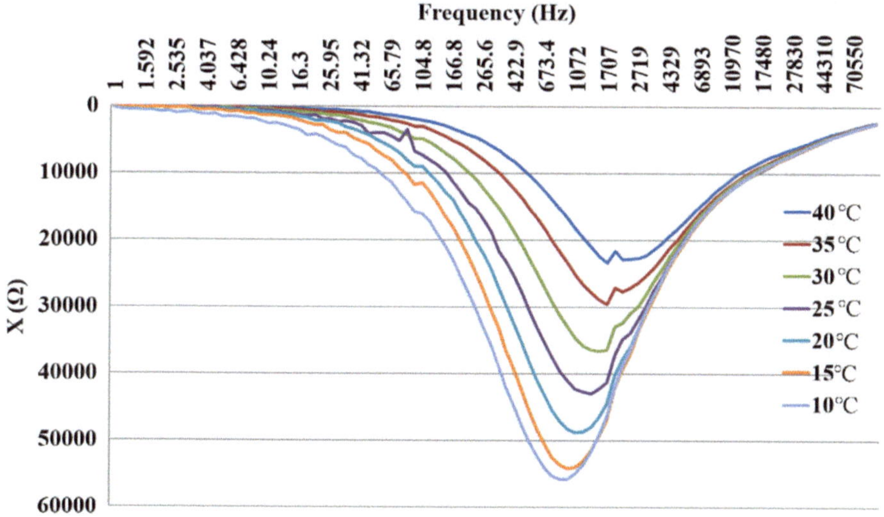

Fig. 4.14 Imaginary part of impedance versus frequency under variable temperature

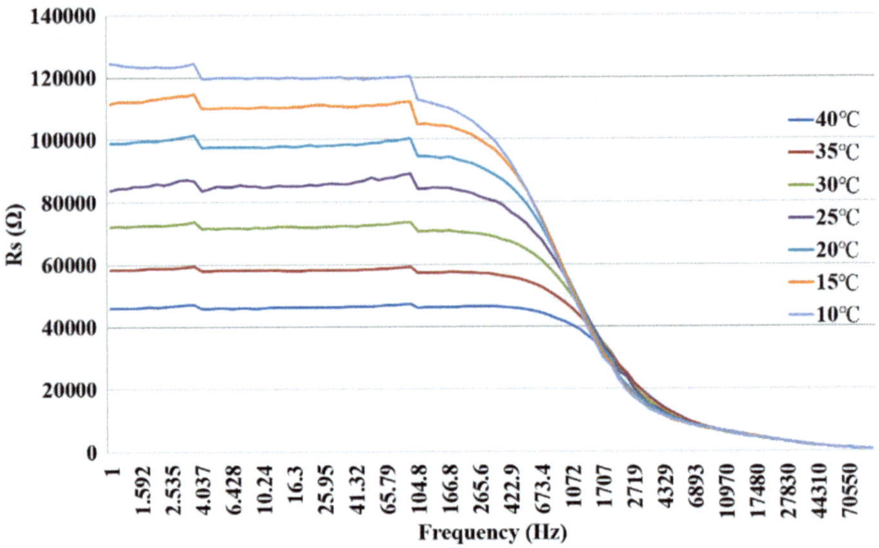

Fig. 4.15 The real part of impedance and frequency

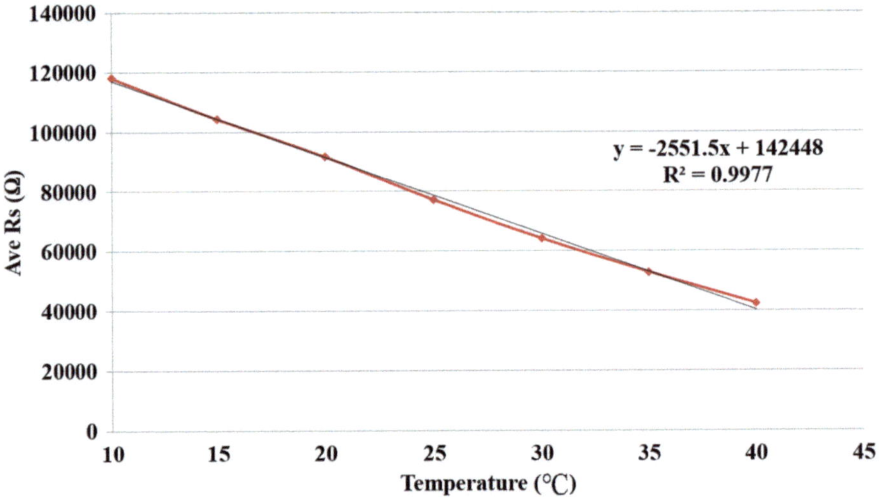

Fig. 4.16 Resistance as a function of temperature at a frequency of 122.5 Hz

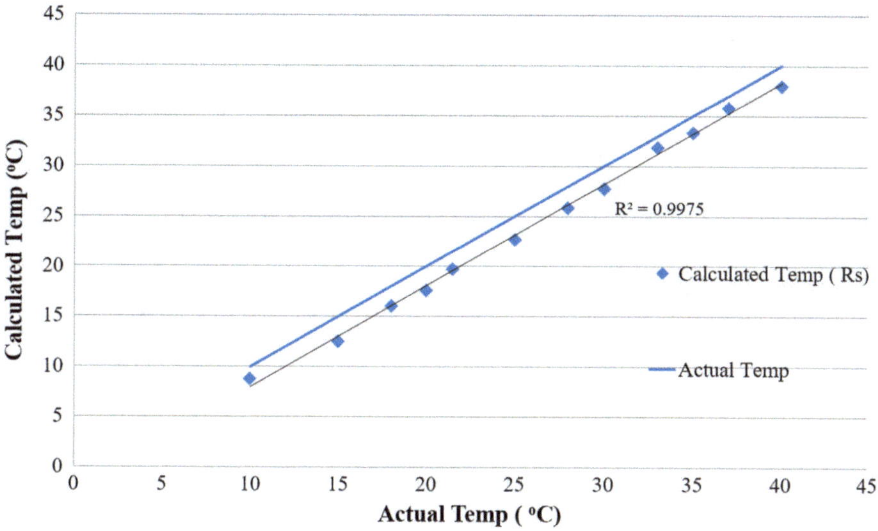

Fig. 4.17 Comparison between the actual temperature and the measured temperature

impedance measurements are taken at the same frequency (122.5 Hz). Therefore, the temperature can be measured by the following equation:

$$T(°C) = \frac{R_T - R_{20}}{slope} + T_{20} \tag{4.8}$$

where, R_T is the measured resistance value, R_{20} is the resistance value at 20 °C (which is the reference value), and T_{20} is the reference temperature at 20 °C. Figure 4.17 displays the correlation between the actual temperature and the calculated temperature. The calculated temperature is well correlated with the resistance ($R^2 = 0.99$) part.

4.4.2 Nitrate Measurement and Standard Equation Development

The sensing system has been used to detect the concentration of nitrate in water. In the laboratory, samples of different concentrations of ammonium nitrate have been prepared by a serial dilution method with the concentration varying from 0.1 to 0.5 mg/L. The impedance of the sensor is measured using electrical impedance spectroscopy technique, and the results are plotted in the Nyquist plot (Fig. 4.18) when the frequency is restricted to 100 kHz. The impedance of the sensor reduces with increased nitrate concentration. Figures 4.19 and 4.20 show the real part and

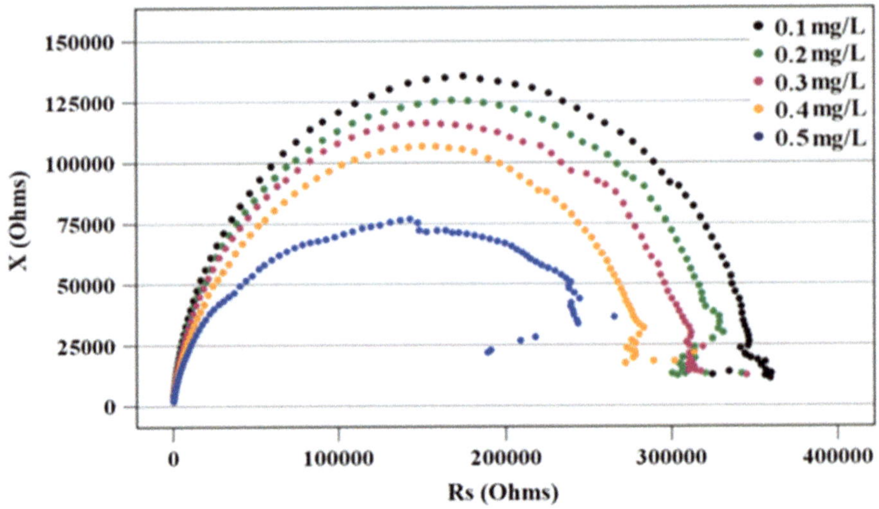

Fig. 4.18 Nyquist plot for ammonium nitrate (NH_4NO_3) at different concentrations

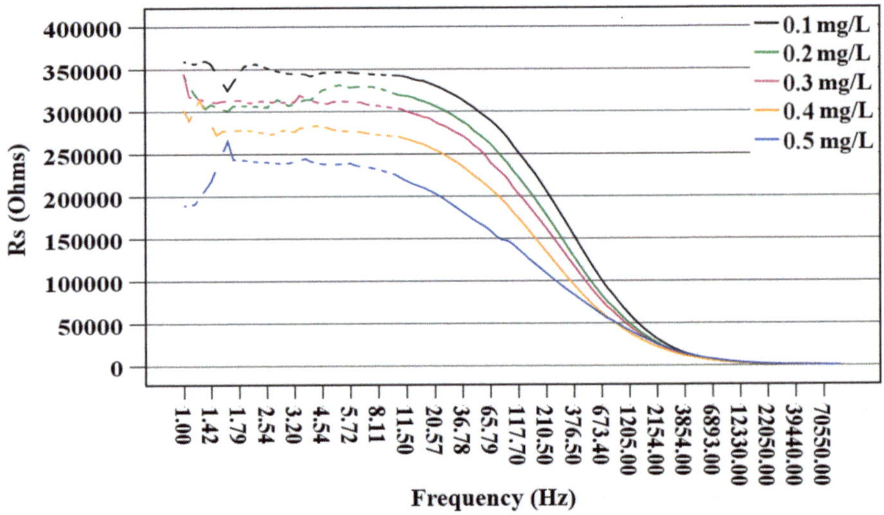

Fig. 4.19 The real part of impedance and frequency

imaginary part of impedance as a function of operating frequencies, for different nitrate concentrations.

The real part of impedance shows a significant change in their value for different concentrations at low frequencies. The imaginary parts do not show much change at low frequencies. Since ionic components play a significant role, the real part of

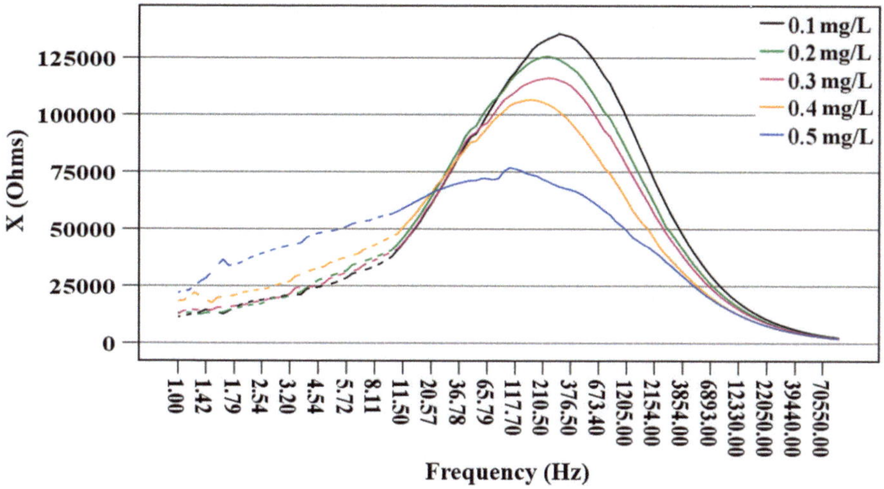

Fig. 4.20 Imaginary part of impedance as a function of frequency

impedance is used to determine the nitrate concentration. The following equation from [20] was derived from the standard curve:

$$C = \frac{R - R_{0.3}}{-115{,}543} + C_{0.3} \tag{4.9}$$

where R is the real impedance of the sensor using water samples, $R_{0.3}$ is the real part of the impedance at 0.3 mg/L, and $C_{0.3}$ represent the concentration of 0.3 mg/L, which is assumed as a reference. Substituting the value of $R_{0.3}$ and $C_{0.3}$ into Eq. (4.9), the concentration was calculated by:

$$C = \frac{R - 48{,}056.78}{-115{,}543} + 0.3 \tag{4.10}$$

Since the sensor is very sensitive to temperature change, the relationship between the resistance (R) and the temperature (T) can be used as a correction factor for the changing rate of R on changing T. Therefore, using the correction factor, the R_{acutal} is modified by Eq. (4.11):

$$R_{actual} = R + \alpha \times (T - T_{20}) \tag{4.11}$$

where T_{20} is the temperature at 20 °C. R is replaced with R_{actual} in Eq. (4.10) to calculate the final computational formula of nitrate-N concentration. Therefore, the standard formula to calculate concentration with a correction factor for the Parylene coated sensor can be represented by Eq. (4.12):

Table 4.1 Comparison of nitrate concentrations measured by the sensor system and the spectrophotometric laboratory method

Lab number	Description	Spectrophotometric method (mg/L)	Sensor system (mg/L)
S202	Stream entering farm	0.50	0.53
S204	Stream entering wetland	0.49	0.53
S206	Stream paddock 9	0.47	0.51
S207	Stream paddock 10	0.47	0.52
S209	Stream exiting farm	0.50	0.55

$$C = \frac{R_{actual} - 48{,}056.78}{-115{,}543} + 0.3 \quad (4.12)$$

This equation has been used to estimate the nitrate concentration in an unknown sample.

4.4.3 Stream Water Testing

The next experiment was conducted with stream water samples using the developed sensing system. The water was collected from different streams and was analysed for nitrate using the spectrophotometric method in the laboratory. The developed standard formula in Eq. (4.12) has been used to calculate the concentration using the designed system. Table 4.1 shows the results from the designed system and the laboratory spectrophotometric method. The spectrophotometric method was taken as the standard laboratory measurement. The stream water samples not only contain nitrate, but also contain phosphates, ammonium, and other mineral salts. This explains the differences between these results. Despite this, the maximum error is less than 10% and is low, the reasons being (i) the calibration of the sensor was done based on nitrate measurement and (ii) the influence of other ions is not strong concerning the nitrate ions.

4.4.4 Data in Cloud Server

In the fourth experiment, the interval time of pump in and pump out of the sample water was 10 min and initially deionised water was pumped out and then 0.5 mg/L Ammonium Nitrate (NH_4NO_3) solution was pumped out for the rest of the period. The microcontroller collects the data and sends the final measured data to the Thingspeak server. Figure 4.21 shows the real-time data which has been sent from the developed smart sensing system.

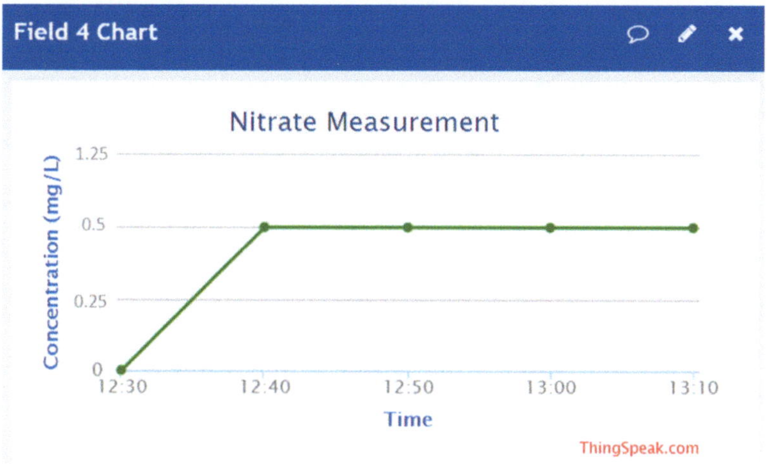

Fig. 4.21 Final data to the cloud server

4.4.5 Comparison of Impedance Measurement by LCR and the Developed System

The fifth experiment was to compare the impedance measurement using the LCR meter and the developed system. Different known concentrations of Ammonium Nitrate (NH_4NO_3) between 0.01 and 0.5 mg/L were measured by the developed system and the LCR meter.

Figure 4.22 illustrates the test results (NH_4NO_3) from the designed system and LCR meter. When the concentration is increased in the sample, the real part of the impedance decreased accordingly. The corresponding impedance calculation was done inside the microcontroller, and the results of the impedance calculation from the system were very close to the impedance measured by the LCR meter. The designed system showed an excellent linear relation with $R^2 = 0.99$ which is very similar to the LCR impedance measurement.

Figure 4.23 shows the comparison of the phase angle between the LCR and the sensor system measurement. During the impedance measurement, the phase angle of the different concentrations was measured. It should be noted that the system's measured phase angles are similar to the LCR measurement and the range of change of phase angle of the sensor system, is also similar to that of LCR measurement. The LCR meter has been used to measure the impedance of the Interdigital sensor for different applications. Therefore, it is essential to compare both results to evaluate the system's performance. The system is capable of running the sensor and recording the data continuously.

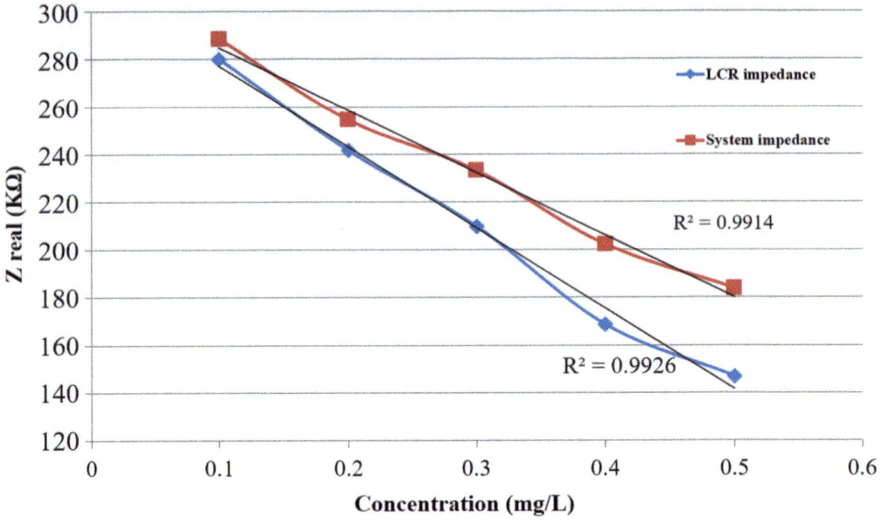

Fig. 4.22 Comparison of the real part of impedance: by LCR and developed system

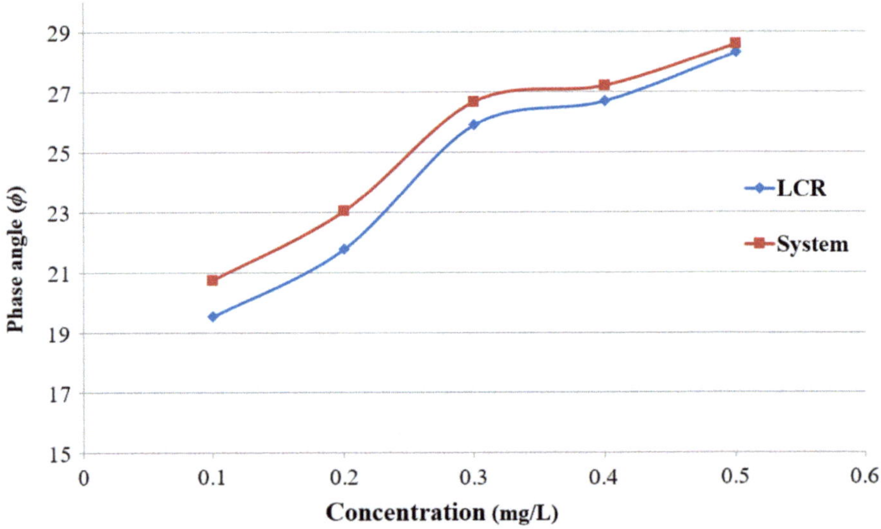

Fig. 4.23 Comparison of phase angle by LCR and developed system

Table 4.2 Comparison of nitrate measurement before and after the temperature compensation

Sample number	Sensor system		Spectrophotometric method (mg/L)
	Before compensation (mg/L)	After compensation (mg/L)	
1	0.61	0.53	0.50
2	0.59	0.51	0.49
3	0.61	0.50	0.48
4	0.41	0.35	0.33
5	0.43	0.30	0.32

4.4.6 *Improvement of Temperature Compensation*

The last experiment was done to demonstrate the usefulness of the temperature compensation in the developed system. The temperature was maintained at 30 °C of the sample water. The concentration of nitrate was measured by the developed sensing system and laboratory spectrophotometric method. Table 4.2 compares nitrate concentrations measured before and after temperature compensation and calculated using Eqs. 4.10 and 4.12, respectively. The error rate was higher (more than 20%) when the temperature change was not compensated in the developed system (Table 4.2).

4.5 Chapter Summary

A temperature compensated interdigital capacitive sensor has been developed in the current chapter to measure nitrate at low concentrations. A portable, novel sensing system has been developed that could be used on-site as a stand-alone device, as well as IoT-based remote monitoring smart sensor node, to measure nitrate concentration in surface and groundwater. Electrochemical Impedance Spectroscopy was employed to detect and display nitrate concentrations, by evaluating the impedance change read by the interdigital transducer immersed in the surface water samples. The test samples were evaluated by commercial equipment (LCR meter) and the designed system. These results were also validated using standard laboratory techniques to assess nitrate concentrations in water samples. The designed system showed an excellent linear relationship between the measured nitrate concentrations (ranged from 0.01 to 0.5 mg/L) to those measured by the commercial equipment in the collected water samples. However, the current system has the potential to be used to estimate nitrate concentrations in water samples, in real-time.

References

1. A.I. Zia et al., Electrochemical impedance spectroscopy based MEMS sensors for phthalates detection in water and juices. J. Phys.: Conf. Ser. **439**, 012026 (2013)
2. M. Yunus, S. Mukhopadhyay, K. Jayasundera, A novel planar interdigital sensor for environmental monitoring, in *Sensors, 2009 IEEE* (IEEE, 2009), pp. 105–110
3. A. Prodic, D. Maksimovic, R.W. Erickson, Design and implementation of a digital PWM controller for a high-frequency switching DC-DC power converter, in *Industrial Electronics Society, 2001. IECON'01. The 27th Annual Conference of the IEEE*, vol. 2 (IEEE, 2001), pp. 893–898
4. C.S. Eva Murphy. All about direct digital synthesis. Analog Dialogue. http://www.analog.com/library/analogdialogue/archives/38-08/dds.pdf
5. "PWM Sine Wave Generation." http://web.csulb.edu/~hill/ee470/Lab%202d%20-%20Sine_Wave_Generator.pdf
6. L. Liao, W. Jin, R. Pavel, Enhanced restricted boltzmann machine with prognosability regularization for prognostics and health assessment. IEEE Trans. Industr. Electron. **63**(11), 7076–7083 (2016)
7. S. Pirbhulal, H. Zhang, S.C. Mukhopadhyay, W. Wu Y.-T. Zhang, Heart-beats based biometric random binary sequences generation to secure wireless body sensor networks. IEEE Trans. Biomed. Eng. (2019)
8. S. Pirbhulal, O.W. Samuel, W. Wu, A.K. Sangaiah, G. Li, A joint resource-aware and medical data security framework for wearable healthcare systems. Future Gener. Comput. Syst. **95**(19), 382–391 (2019)
9. A.H. Sodhro, S. Pirbhulal, Artificial intelligence driven mechanism for edge computing based industrial applications. IEEE Trans. Industr. Inf. **19**(1), 1–8 (2019)
10. S. Pirbhulal, P. Shang, W. Wu, O.W. Samuel, G. Li, Fuzzy vault-based biometric security method for tele-health monitoring systems. Comput. Electr. Eng. **71**(1), 546–557 (2018)
11. S. Pirbhulal, P. Shang, W. Wu, A.K. Sangaiah, O.W. Samuel, G. Li, Fuzzy vault-based biometric security method for tele-health monitoring systems. Comput. Electr. Eng. **71**, 546–557 (2018)
12. W. Wu, S. Pirbhulal, G. Li, Adaptive computing-based biometric security for intelligent medical applications. Neural Comput. Appl. pp. 1–10 (2018)
13. S. Pirbhulal, H. Zhang, S.C. Mukhopadhyay, W. Wu Y.-T. Zhang, An efficient biometric-based algorithm using heart rate variability for securing body sensor networks. Sensors, SCI, **15**, 15067–15089 (2015)
14. S. Pirbhulal, H. Zhang, W. Wu, Y.-T. Zhang, Analysis of Efficient Biometric Index using Heart Rate Variability for Remote Monitoring of Obstructive Sleep Apnea. Neuropsychiatry, SCI (2017)
15. L. Da Xu, W. He, S. Li, Internet of things in industries: a survey. IEEE Trans. Industr. Inf. **10**(4), 2233–2243 (2014)
16. L. Li, H. Xiaoguang, C. Ke, H. Ketai, The applications of WiFi-based wireless sensor network in internet of things and smart grid, in *2011 6th IEEE Conference on Industrial Electronics and Applications (ICIEA)* (IEEE, 2011), pp. 789–793
17. Thingspeak: The open data platform for the Internet of Things. https://thingspeak.com/
18. T. Berners-Lee, R. Fielding, H. Frystyk, Hypertext transfer protocol–HTTP/1.0, 2070-1721 (1996)
19. M.E.E. Alahi, X. Li, S. Mukhopadhyay, L. Burkitt, Application of practical nitrate sensor based on electrochemical impedance spectroscopy, in *Sensors for Everyday Life: Environmental and Food Engineering*, ed. by S.C. Mukhopadhyay, O.A. Postolache, K.P. Jayasundera, A.K. Swain (Springer International Publishing, Cham, 2017), pp. 109–136
20. M.E. Alahi, L. Xie, A.I. Zia, S. Mukhopadhyay, L. Burkitt, Practical nitrate sensor based on electrochemical impedance measurement, in *Instrumentation and Measurement Technology Conference Proceedings (I2MTC), 2016 IEEE International* (IEEE, 2016), pp. 1–6

Chapter 5
Graphite/PDMS Capacitive Sensor for Nitrate Measurement

5.1 Introduction

This chapter describes the fabrication and implementation of the second type of sensor patches developed from Aluminium (Al) and Polyethylene Terephthalate (PET). One distinct difference between the structure of these sensor patches is their formation from a single raw material. Due to this attribute, these sensor patches are different regarding performance from other sensors. It also helped in curbing the complexity in the construction process of the sensor prototypes and so decreasing the overall cost of fabrication. PET is one of the conventional polymers used as a substrate to develop flexible sensors for different strain [1, 2] and pressure [3, 4] sensing applications. The absence of any post-processing steps, high flexibility, and smooth cut edges are some of the advantages of this polymer [5, 6]. Aluminium particles also have certain advantages which make Aluminium a viable option for the electrode material [7]. Some of them are high corrosion resistance, high electrical and thermal conductivity, and flexibility [8]. Similarly, to the CNT-PDMS sensors, laser cutting was used to develop the electrodes of these sensor patches.

5.2 Fabrication of the Printed Sensors

The fabrication of the sensor patches was done at fixed temperature and humidity conditions (temperature: 22 °C, RH: 50%). Figure 5.1 shows the schematic diagram of the fabrication steps of the CNT-PDMS based sensor. The Polydimethylsiloxane (PDMS) was cast on a Poly (methyl methacrylate) PMMA template to form the substrate for the sensor patch. PDMS was used to form the sensor patch due to its low-cost, low Young's modulus (E), and hydrophobic nature. PMMA was used due to its non-toxicity and proper adherence with the cured PDMS. Then, a layer of nanocomposite (NC) formed by mixing Multi-Walled Carbon Nanotubes (MWCNTs) and PDMS was cast on the cured PDMS. MWCNTs were considered as the

Md E. E. Alahi and S. C. Mukhopadhyay, *Smart Nitrate Sensor*, Smart Sensors, Measurement and Instrumentation 35, https://doi.org/10.1007/978-3-030-20095-4_5

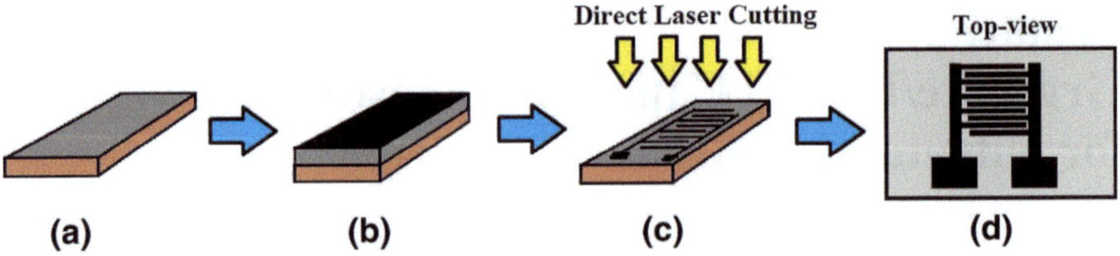

Fig. 5.1 Schematic diagram of the CNT-PDMS based sensor. **a** PDMS was cast on a PMMA template. **b** A layer of nanocomposite (NC) layer was cast on top of the cured PDMS. **c** The cured NC layer was laser cut to form the electrodes. **d** Final product used as a sensor

conductive material due to their high electrical conductivity, flexibility, high tensile strength and resistance towards a wide range of temperatures. The curing of the NC layer was followed by the laser cutting of the top layer to form the electrodes of the sensor patch. The individual fabrication steps are shown in Fig. 5.2. PDMS (SYLGARD ® 184, Silicon Elastomer Base) was formed by mixing a ratio of 10:1 between the base elastomer and curing agent respectively. The height of the PDMS cast on the PMMA was adjusted by a casting knife (SHEEN, 1117/1000 mm) to around 1000 microns. The sample was then desiccated for 2 h to remove any trapped air bubbles. After curing the sample at 80 °C for 8 h, a layer of NC formed by mixing MWCNTs (Aldrich, 773840-100G) and PDMS was cast on top of it. The weight value of the MWCNTs was optimised to have a trade-off between the flexibility of the sensor patch and the conductivity of the electrodes. The final optimised value of the CNTs chosen to form the NC was four wt%. Figure 5.3a shows a Scanning Electron Microscope (SEM) image of the optimised CNT in NC. The black regions in the image represent the CNTs while the white spots define the PDMS in NC. The casting knife again adjusted the height of the NC layer to around 600 microns. The sample was then again desiccated followed by its curing at 80 °C for 8 h to form the top layer of the sensor patch. Then the sample was taken for laser cutting to scan off the top part except for the design electrodes. A Universal Laser Systems (Model: OLS 6.75, CO_2 laser system, laser spot diameter: 150 microns) was used for laser cutting on the sample to form the electrodes. The design of the electrodes was done with the help of Corel DRAW as it was the design software available with the Universal Laser associated with the laser cutting system.

Three parameters, namely power (W), speed (m/min) and z-axis (mm) were optimised to achieve an optimal cut of reasonable quality. The power is defined as the power exerted by the laser beam. The speed was used to control the movement of the laser nozzle over the sample. Z-axis changed the height of the laser platform to adjust the focal point of the laser beam on the sample. The optimised values of these three parameters were 24 W, 70 m/min and 1 mm. Figure 5.3b shows the SEM image of the finished product. The front and rear view along with the dimensions of the sensor patch is shown in Fig. 5.4. There were four pairs of interdigitated electrodes with a thickness of 200 microns for each finger and a gap of 100 microns between two consecutive fingers. The length and width of the sensing area of the patch were

Fig. 5.2 Individual fabrication steps followed in developing the CNT-PDMS sensor

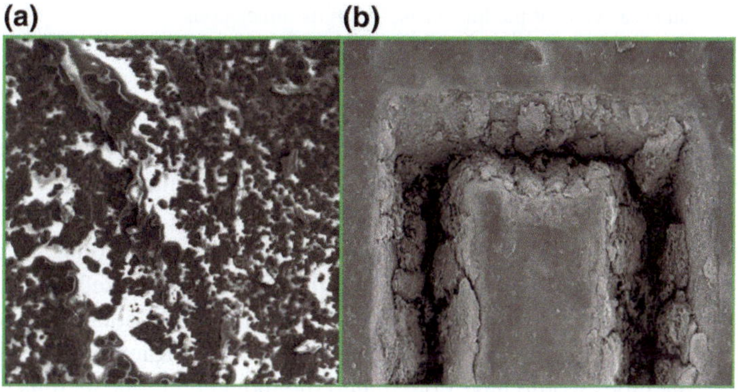

Fig. 5.3 SEM image of the **a** CNT-PDMS mixture with the optimised CNT wt% and **b** formed a sensor patch

5 and 4.5 mm respectively. Two of the advantages of these sensor patches were their enhanced mechanical properties and the corrosion-resistant nature of the electrodes.

The schematic diagram of the second type of sensor is given in Fig. 5.5. The electrodes of these sensor patches were developed from low-cost polyimide (PI) films by photo-thermal induction [9]. The laser cutting of the PI films was followed by transfer of the formed conductive material onto Kapton tapes by manually compressing the tapes over the induced graphene. The Kapton tapes with the transferred graphene were then used as sensor patches for experimental purposes. Figure 5.6 shows the individual steps of fabrication of the graphene sensor. After the design of the electrodes was provided to the laser cutting system, the PI films, of a thickness of around 100 microns, were taken to the platform for laser writing purposes. The PI films were attached to a glass substrate to forbid movement during the induction process. The three laser parameters (power, speed, and z-axis) were again optimised to produce the conductive material from the PI films. The optimised values were 9 W, 70 m/min and 1 mm for the power, speed and z-axis respectively. The sp^3 hybridised

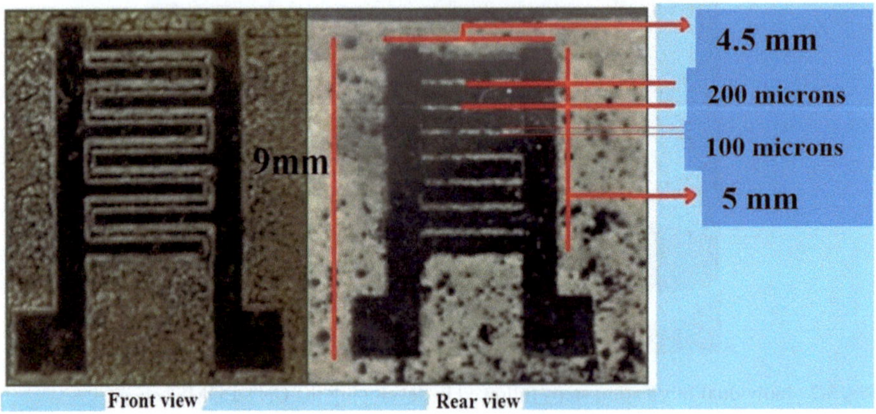

Fig. 5.4 Front and rear view of the final sensor with its dimensions

carbon atoms of the PI films were photo-thermally converted to sp^2 hybridised carbon atoms of graphene. After the formation of graphene, Kapton tapes were laid over the designed electrodes to transfer the induced conductive material. The transfer was done carefully by initially applying manual pressure over the sensing area of the patches, followed by transferring the bonding pads of the electrodes. Even though the PI film used to generate the graphene and the Kapton tapes used as a substrate of the sensor patches are of the same material, there are two reasons the laser writing was done on the PI film instead of doing it directly on the Kapton tapes. Firstly, the stickiness of the tapes would have coagulated the induced graphene, thus hampering the design. Secondly, the conductivity of the induced graphene would have been affected by the Kapton tapes. The conductivity of the conductive material was not much affected by the transfer as the difference was less than 20 mS/m between the induced graphene and the transferred graphene. Figure 5.7 shows SEM images of the top and side views of the electrodes of the electrode on the Kapton tapes. It is seen that the transfer came off clean with the edges most of the fingers parallel to each other. Figure 5.8a shows the image of the sensor patch with its dimensions. Six pairs of electrode fingers were formed with each one having a length and width of 500 microns and 100 microns respectively. The interdigital distance between two consecutive electrode fingers was around 100 microns. Figure 5.8b shows the equivalent circuit of the sensor patch. V_{in} is the input voltage, and I_{in} is the current which is flowing through the electrodes. R_p and C_p are the electrical components, which are formed due to the influence of the electric field applied to the electrodes. R_p is the total resistance due to the conductive properties of the sample and C_p is the capacitance which formed due to dielectric properties of the medium of sample [10].

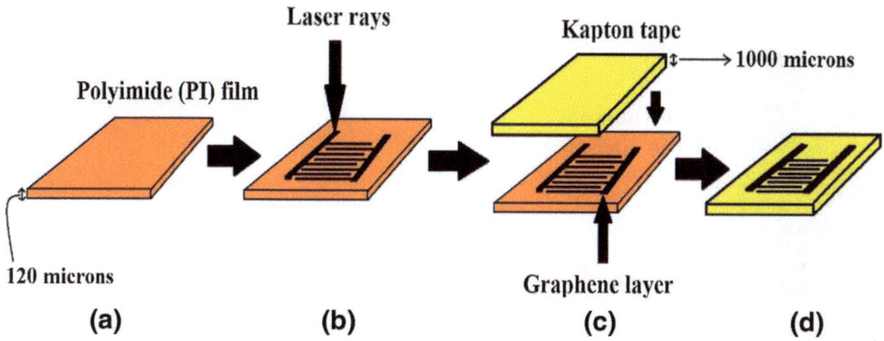

Fig. 5.5 Schematic diagram of the graphene sensor. The polyimide (PI) film was taken (**a**) for laser writing on it (**b**). The induced graphene electrodes (**c**) were transferred to the Kapton tape to form the sensor patch (**d**)

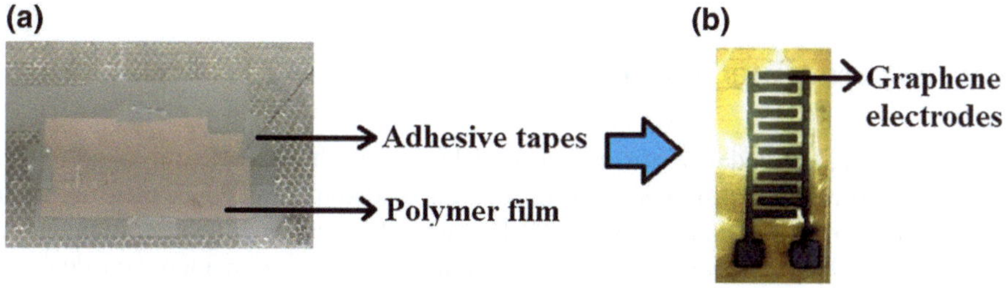

Fig. 5.6 The polymer film was added on a glass substrate **a** to induce graphene, which was transferred on Kapton tapes **b** to use it as electrodes on a sensor patch

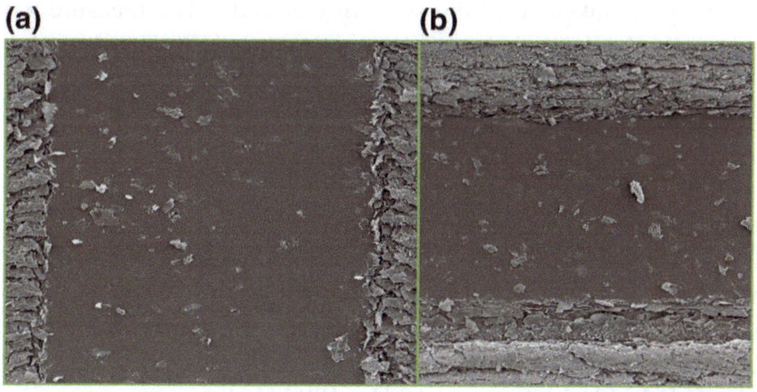

Fig. 5.7 SEM image of the **a** side view and **b** top view of the transferred graphene on the Kapton tape

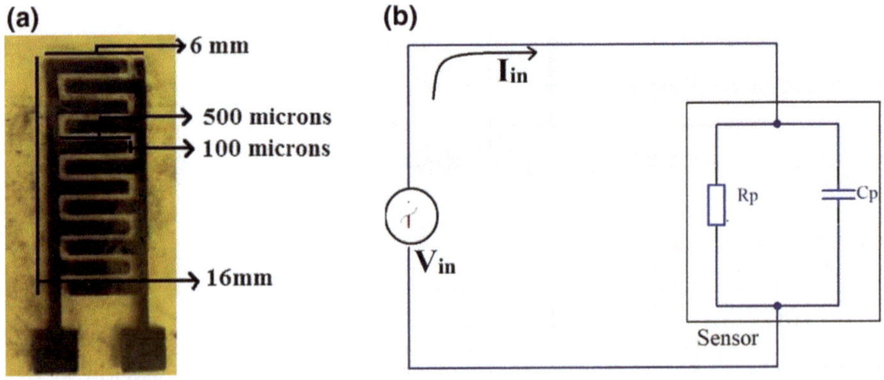

Fig. 5.8 **a** Final product along with its dimensions that were used as a sensor patch, **b** The equivalent circuit of the Interdigital sensor

5.3 Experimental Setup

Electrochemical Impedance Spectroscopy (EIS) is a highly sensitive method for an uncertain and variable system in equilibrium, and its instant impedance measurement is required for non-stationary systems. Different methods are available for impedance measurement, but the Frequency Response Analyser (FRA) is considered the de facto standard for EIS measurement. An FRA requires a single input sinusoidal signal with an amplitude of 5–15 mV. The frequency of the signal sweeps in a certain range on a direct current bias voltage. The signal is applied to on the working electrode, and a resulting voltage is taken from the sensing electrode. The measurement process is done in a particular frequency range to get a complete impedance profile. The following can be used to specify the impedance of a system:

$$Z = R + jX \tag{5.1}$$

where Z is the impedance (Ω); R is the resistance (Ω), which is also the real part of the impedance; X is the reactance (Ω), which is the imaginary part of the impedance.

The impedance profile data can be represented graphically as a Bode plot, and a Nyquist plot also called a Cole-Cole plot. The plot also reflects the electrochemical procedures taking place at the electrode-electrolyte edge. A clamp connects the sensor with the HIOKI IM 3536 LCR meter. 10 Hz–100 kHz frequency sweeping was used to profile the sensor for EIS measurement. Standard temperature and humidity were maintained during measurement. Deionised water was used as a control solution. A standard nitrate-N solution was used to prepare the working solution for different concentrations. The average pH of the samples was 6.60. Initially, the sensors were characterised in the air to extract the experimental reference curve. Then the deionised water and sample water were used to continue the experiments. Figure 5.9 shows the data-acquisition laboratory setup used for EIS measurements.

Fig. 5.9 Experimental setup for nitrate measurement

5.4 Comparative Analysis of Two Different Sensors

10-ppm Nitrate-N sample water was taken, and the real part of the impedance was measured for different sensors. The real part of the impedance of the deionised water was also measured. The temperature and humidity of the waters were identical due to the same laboratory environment. The pH of the water was also the same during the measurement. The sensor's response was calculated by using Eq. 5.2.

$$\boldsymbol{Sensor's\, response}(\%) = \frac{\boldsymbol{R}(\boldsymbol{milliq}) - \boldsymbol{R}(\boldsymbol{sample})}{\boldsymbol{R}(\boldsymbol{milliq})} \times 100 \qquad (5.2)$$

where, R indicates the real part of the impedance of the sensors.

5.5 Temperature and Nitrate-N Measurement

The graphene sensor was first used to measure temperature changes in water. Temperature changes the mobility of ions in water; therefore, it is essential to measure the changing behaviour of the graphene sensor at different temperatures. An experimental setup similar to Fig. 5.9 was used to measure the temperature. The GEX MS 7-H550 Digital Hotplate, mercury thermometer, LCR meter, and the computer was used for data acquisition. The thermometer was immersed inside the deionised water to measure the temperature instantly. The sensing surface of the graphene sensor was immersed inside the water, and the frequency was swapped from 10 Hz–100 kHz to characterise the sensor under different temperatures. 1, 10, 30, 50 and 70-ppm standard nitrate-N solutions were taken for another experiment 100 ml in five beakers. 100 ml of deionised water was also taken as a control solution. The sensor was immersed in the sample waters until the completion of the measurements. The real and imaginary impedance were taken by the LCR meter. The measurement was taken five times, and an average was presented in the final graph.

5.6 IoT-Enabled Smart Sensing System

Internet of Things (IoT) enabled smart sensing system is proposed to collect the sensor data for nitrate-N measurement. An AD5933 [11] was used as an impedance analyser to measure the real part of the impedance of the graphene sensor. Arduino Uno Wi-Fi [12] is used as a master microcontroller to collect the impedance data from the impedance analyser. It has an integrated Wi-Fi module, which would help to send data wirelessly to a cloud server. Thingspeak [13] is the IoT-based free web server to store the measured data and show it in real-time. The Arduino Ciao [14] library was used to transfer the data to the designated private channel in Thingspeak. Ciao is a library which is capable of interfacing with system resources and communicating with the most common protocols such as (MQTT, XMPP, HTTP, SMTP, etc.). HTTP POST was used to send the measured concentration to the *ThingSpeak* cloud server. Security is an essential issue in IoT research due to the large scale of the objects and the heterogeneity. It will be discussed in a future research article when the actual distribution network will be installed to monitor in real time.

The AD5933 is the impedance analyser where it measures the impedance and the phase shift of the sensor. Before starting the measurement, the AD5933 had to be calibrated to get the gain of the impedance analyser. An ADG849 is used as a switch from a calibration resistor to the sensor, and the impedance is calculated by using that calibrated gain. The phase shift is also calculated from the impedance analyzer. Frequency sweeping is not required as the final measurement depends on the single frequency. The I^2C protocol is used to communicate to the impedance analyzer to extract the real part of the impedance from the impedance analyzer and store that in the main microcontroller. A data processing algorithm is used to convert that real part of the impedance into meaningful temperature and nitrate-N concentration values.

Finally, the temperature and nitrate-N data are sent to the IoT-based cloud server to store for further processing. Some of the sample water, which was collected from different locations, is measured by the developed sensing system. Figures 5.10 and 5.11 show the block diagram and software flow of the developed sensing system. Arduino Sketch was used by writing the programming code. Initially, the system starts, and the microcontroller is initialised the setup function to establish the rest connector, provided with the cloud-server API (Application Programming Interface) number which will be necessary to send the measured data to the designated IoT server. The impedance analyser is used to measure the phase and real part of the impedance of the sensor. The developed calibration standard for temperature measurement was used to measure the temperature of the deionised water. After that, the nitrate-N concentration was calculated, with temperature compensation. The system communicates with the IoT cloud server to send the measured data. Finally, the microcontroller goes into sleep mode to save power.

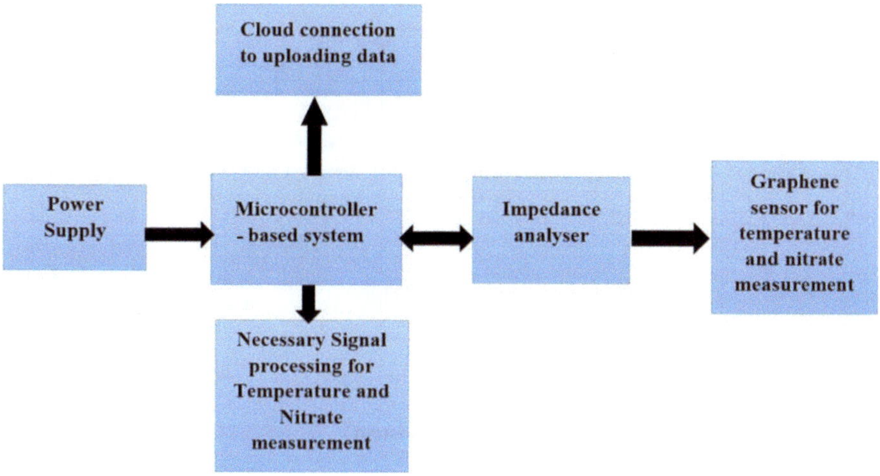

Fig. 5.10 Block diagram of the smart sensing system

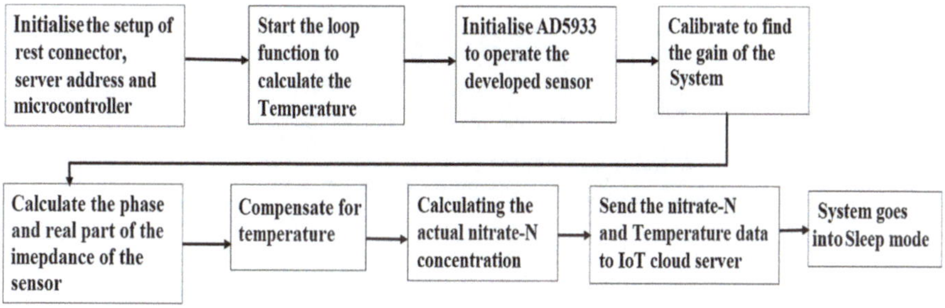

Fig. 5.11 Software flow of the individual steps of the operating of the IoT-based system to calculate and transmission of the nitrate concentration to the cloud server

Figures 5.12 and 5.13 show a schematic diagram of the conditioning circuit and the prototype of the sensing system respectively. Circuit maker [15] was used to draw the circuit diagram of the sensing system.

5.7 Results and Discussion

5.7.1 Comparative Analysis

The sensor's response is calculated for two different sensors and plotted in Fig. 5.14. It is seen from Fig. 5.14 that the Graphene sensor has a higher sensitivity compared to CNT PDMS sensor. The materials of the electrode have an impact on this sensitivity curve. The graphene has an excellent electrical conductivity compared to

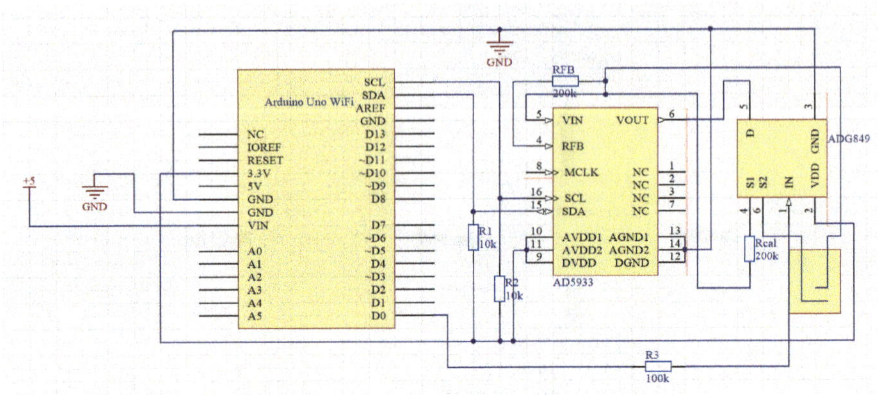

Fig. 5.12 Schematic diagram of the smart sensing system

Fig. 5.13 Prototype of the smart sensing system

PDMS+CNT material. The electrical intensity, which starts from the positive electrode, can bulge through the sample water to the negative electrode. Therefore, the graphene sensor is showing a better response compared to the other sensor. The response of the graphene sensor is also stable in all the frequency range. In the subsequent sections, all the experiments were done by using the graphene sensor due to the higher response compared to the CNT PDMS sensor in this study.

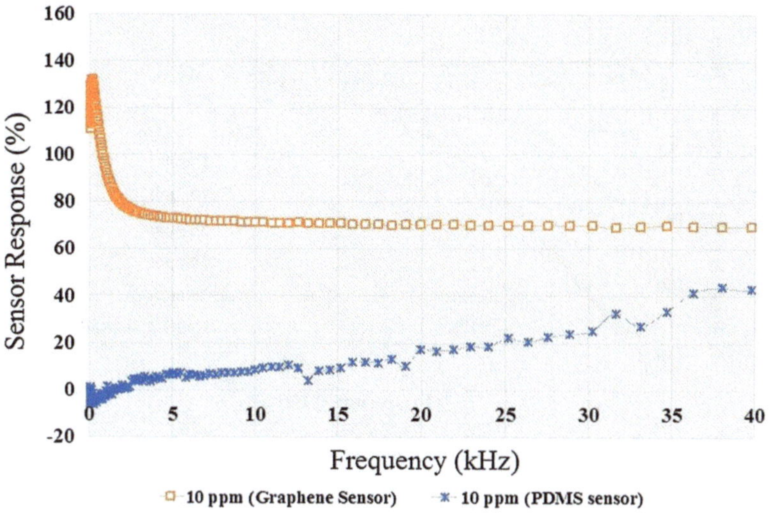

Fig. 5.14 Comparison of the sensitivity of different sensors

5.7.2 Nitrate Measurement

Figure 5.15 presents the real impedance for different concentrations with a frequency range from 1 Hz to 100 kHz. It is seen that the real part of the impedance is changing due to the different concentrations of nitrate-N in the sample water. Due to the presence of nitrate-N ions in the water, the impedance profile is changed for different concentrations due to the properties of the interdigital graphene sensor. It is observed that 1200–1700 Hz is the sensitive region for different concentrations. Among the real and imaginary parts of the impedance of the graphene sensor, the real part of the impedance shows the most change between the sample nitrate-N concentrations.

5.7.3 Temperature Measurement

The dielectric properties of the deionised water were measured by this described method. The EIS measurement technique was used to measure the real and imaginary parts of the impedance of the sensor. The real part gives a more significant change compared to imaginary impedance. The temperature was varied from 7 to 50 °C, and the corresponding real part of the impedance was plotted in Fig. 5.16. From the linear regression analysis, it is seen that the temperature is well correlated with the real impedance ($R^2 = 0.99$). Therefore, the temperature can be calculated from Eq. 5.3:

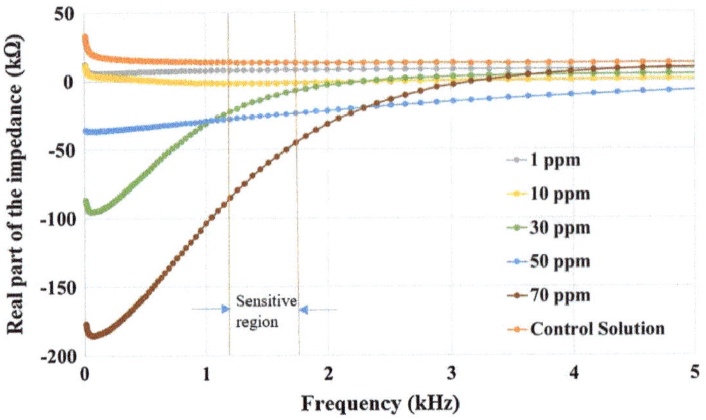

Fig. 5.15 The change of the real part of the impedance concerning frequency

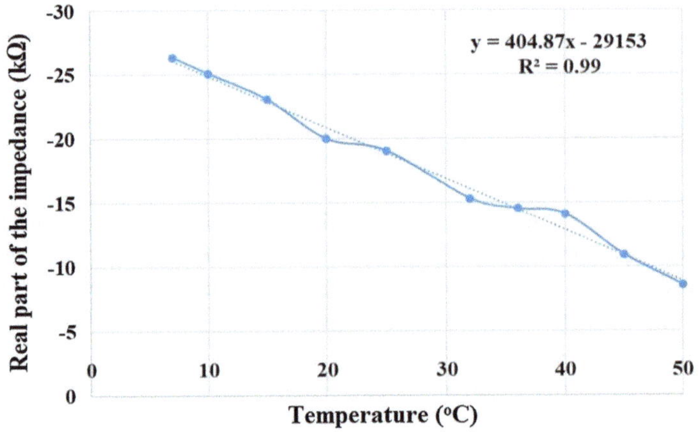

Fig. 5.16 Real part of the impedance as a function of temperature

$$T(^{\circ}C) = \frac{(R_T + 29153)}{404.87} \tag{5.3}$$

where, R_T is the measured real part of the impedance for a certain temperature and T is the calculated temperature.

The slope of the straight line indicates the change of the real part of the impedance per unit change of temperature, which is $\alpha = 404.87\ \Omega/\ ^{\circ}C$. It is to be noted that, the operating frequency of the graphene sensor of the temperature measurement was 1650 Hz. Figure 5.17 shows the temperature measured by the sensor and compares it with the actual temperature. It is seen that they are well correlated with each other and $R^2 = 0.99$ which indicates that the sensor can quite accurately measure the temperature of the sample water.

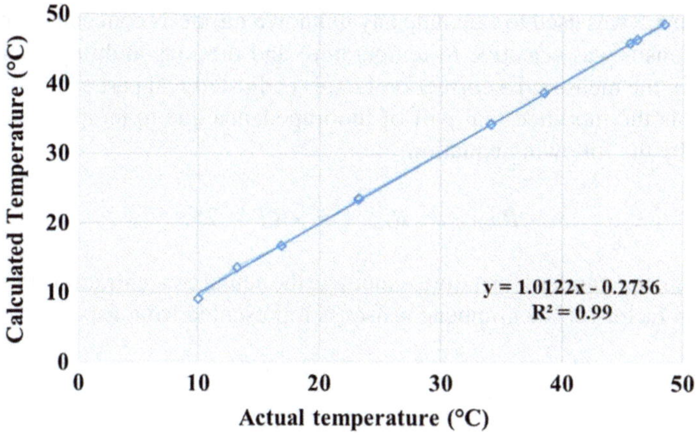

Fig. 5.17 Comparison of actual and measured temperatures

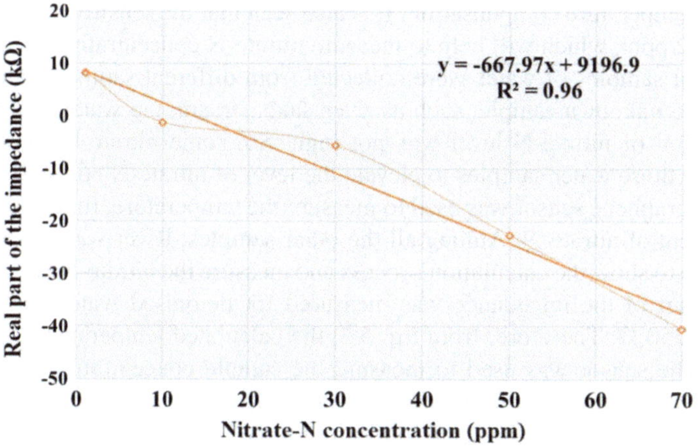

Fig. 5.18 Calibration standard of Nitrate-N concentration (ppm)

1650 Hz was taken as the operating frequency to develop the calibration standard from a standard nitrate-N sample measurement. All the measured concentrations were considered as the x-axis, and the corresponding real part of the impedance was considered as the y-axis. Figure 5.18 shows the final calibration standard for nitrate-N measurement. Linear regression analysis can be calculated by from Eq. 5.4:

$$C = \frac{R_{cal} - 9196.9}{-667.97} \tag{5.4}$$

where C (ppm) is the actual concentration, and R_{cal} (Ω) is a real impedance measured by the graphene sensor.

Equation 5.5 was used to calculate any unknown nitrate-N concentration in water. Since the sensor was sensitive to temperature and the ions mobility changes with temperature, the measured a correction factor α adjusted real part of the impedance. The R_{actual} is the modified real part of the impedance due to temperature and was calculated by the following equation.

$$R_{actual} = R_{cal} + \alpha \times (T - 25) \tag{5.5}$$

Therefore, the standard formula to calculate the actual concentration after applying a correction factor for the graphene sensor is represented from Eq. 5.6.

$$C_{actual} = \frac{R_{actual} - 9196.9}{-667.97} \tag{5.6}$$

where C_{actual} is the final corrected concentration.

The final equation was used to measure any unknown nitrate-N concentration, including temperature compensation. It is also seen that the sensitivity of the sensor is 667.97 Ω/ppm, which will help to measure nitrate-N concentrations accurately.

Different samples of water were collected from different sampling locations to measure the unknown sample, such as river, lake, stream, tap water. Naturally, the concentration of nitrate-N level was not high. So, some nitrate-N samples were added with those water samples to elevate the level of nitrate-N concentration. The fabricated graphene sensor was used to measure the temperature, followed by a final measurement of nitrate-N. Among all the other samples, River water was taken as an example to show the calculation process and measure the nitrate-N concentration. The real part of the impedance was measured for deionised water and gave that $R_T = -19{,}750$ Ω. Therefore, from Eq. 5.3, the calculated temperature is T (°C) = 23.2 °C. The sensor was used to measure the sample concentration, and the real part of the impedance was found, $R_{cal} = -4568$ Ω. Using Eq. 5.5, R_{actual} becomes −5288.67 Ω. Therefore, the final concentration is 21.69 ppm, which comes from that sample water measurement. This result was verified by the laboratory standard method of UV-spectrometry. Other sample waters are measured and compared with the laboratory standard method in Table 5.1. It is seen that the graphene sensor shows outstanding performance compared to the laboratory method. The error rate was less than 5%, which is an acceptable performance of the sensing system and the developed sensor. The error was considered as a measurement error due to the presence of other ions in water. However, the accuracy of the sensor and sensing system was more than 95% and consistent.

The developed sensor is robust and maintains good repeatability. Figure 5.19 shows the reusability performance of the sensor and the sensing system. The sensor can provide an almost identical result in each run of the measurement. Graphene is corrosion free [16] and protects the sensing electrodes from oxidation during the sample measurement, which helps when using the sensor in a repetitive manner. Mechanically it is also robust [17] and shows good performance during measurements.

Table 5.1 Unknown sample measurement (in ppm) compared with laboratory standard method

Sl no.	Sampling number	1st run	2nd run	3rd run	4th run	5th run	Laboratory method
1.	River water	21.69	21.56	21.54	21.63	21.35	21.5
2.	Tap water	5.25	5.1	5.35	5.15	5.75	5.5
3.	Canal water	56.75	56.7	56.6	56.65	56.55	56.5
4.	Stream water	65	65.2	65.1	65.3	65.15	65
5.	River water	32.35	32.65	32.55	32.45	32.62	32.5

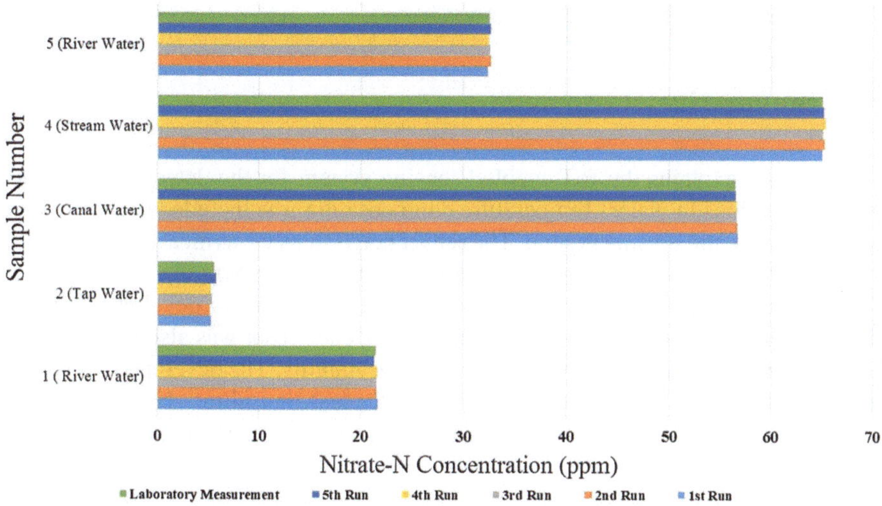

Fig. 5.19 Repeated unknown sample measurements by smart sensing system

Figure 5.20 shows the transfer of the data to the IoT-based cloud server. The concentration of nitrate-N was 25.5 ppm, and the temperature was around 22 °C. The system was used for nearly three hours to monitor the actual data in real time. It was observed that the developed system was consistent in terms of monitoring continuous data at fixed intervals. They have transferred the data simultaneously, which would help to monitor the measured data in real time. There is a certain delay (30 s) from *ThingSpeak* cloud server. The sensing system will also be useful to develop a distributed monitoring system to monitor the temperature and nitrate-N concentration in real time.

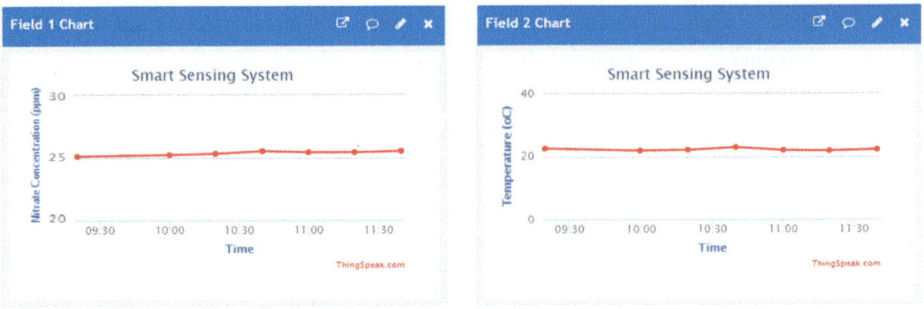

Fig. 5.20 Data transferred to the IoT based web server

5.8 Summarize

This Chapter explains the IoT-enabled sensing system for the detection of nitrate-N concentrations in water. Temperature compensation was included to improve the performance of the sensing system. The sensor performance was measured in terms of its ability to monitor different concentrations of the nitrate-N in the water samples. Due to the low fabrication cost of the sensor, it was easier to develop a low-cost sensing system to monitor the water in real time. The sensor was also robust, which helps to repeat the measurements with good repeatability. The developed sensor and the smart sensing system can be used to monitor the real-time nitrate-N concentrations and develop a low-cost distributed network with sufficient performance.

References

1. J. Singh, H. Chu, J. Abell, R.A. Tripp, Y. Zhao, Flexible and mechanical strain resistant large area SERS active substrates. Nanoscale **4**(11), 3410–3414 (2012)
2. C. Lee, J. Ahn, K.B. Lee, D. Kim, J. Kim, Graphene-based flexible NO2 chemical sensors. Thin Solid Films **520**(16), 5459–5462 (2012)
3. M. Akiyama, Y. Morofuji, T. Kamohara, K. Nishikubo, M. Tsubai, O. Fukuda, N. Ueno, Flexible piezoelectric pressure sensors using oriented aluminum nitride thin films prepared on polyethylene terephthalate films. J. Appl. Phys. **100**(11), 114318 (2006)
4. X. Wang, H. Zhang, R. Yu, L. Dong, D. Peng, A. Zhang, Y. Zhang, H. Liu, C. Pan, Z.L. Wang, Dynamic pressure mapping of personalized handwriting by a flexible sensor matrix based on the mechanoluminescence process. Adv. Mater. **27**(14), 2324–2331 (2015)
5. Laser cutting and engraving of PET. https://www.eurolaser.com/materials/pet-petg/
6. Cutting PET with CO2 laser. http://www.gccworld.com/showcase.php?act=view&no=138
7. A. De La Escosura-Muñiz, M. Espinoza-Castañeda, M. Hasegawa, L. Philippe, A. Merkoçi, Nanoparticles-based nanochannels assembled on a plastic flexible substrate for label-free immunosensing. Nano Res. **8**(4), 1180–1188 (2015)
8. Aluminium–Advantages and Properties of Aluminium. http://www.azom.com/article.aspx?ArticleID=1446
9. J. Lin, Z. Peng, Y. Liu, F. Ruiz-Zepeda, R. Ye, E. L. Samuel, M.J. Yacaman, B. I. Yakobson, J. M. Tour, Laser-induced porous graphene films from commercial polymers. Nat. Commun. 5 (2014)

10. T-T. Ngo, A. Bourjilat, J. Claudel, D. Kourtiche, M. Nadi, Design and realization of a planar interdigital microsensor for biological medium characterization. In *Next Generation Sensors and Systems*, (Springer, 2016), pp. 23–54
11. Devices, A. AD5933: Impedance Analyzer. http://www.analog.com/media/en/technical-documentation/data-sheets/AD5933.pdf (accessed 04/09/2017)
12. Arduino Arduino Uno WiFi. https://store.arduino.cc/usa/arduino-uno-wifi (accessed 4/09/2017)
13. Thingspeak Thingspeak. https://thingspeak.com/ (accessed 26/08/2017)
14. A. Website, Ciao. https://www.arduino.cc/en/Reference/Ciao (accessed 26/08/2017)
15. Circuit maker. https://circuitmaker.com/ (accessed 4/09/2017)
16. D. Cochlin, Graphene paints a corrosion-free future. http://www.manchester.ac.uk/discover/news/graphene-paints-a-corrosion-free-future/. Accessed 28 Aug 2017
17. A.K. Geim, Graphene: status and prospects. Science **324**(5934), 1530–1534 (2009)

Chapter 6
Preparation and Characterization of the Selectivity Material of Nitrate Sensor

6.1 Introduction

The previous chapters explained two different sensors for nitrate detection. One sensor was a MEMS sensor, and another one was a Graphite/PDMS printed sensor. Both of the sensors were low cost and very useful for water quality measurement applications. However, both the sensors did not have any selectivity. Therefore, a selective polymer membrane is developed which can avoid the interference effect in water and provide accurate results. The selective layers worked as coating and exposed to a mixture of ions in water. It interacts with those ions for which the sensor is selective and rejects the other interfering ions. The selective layer of the sensor can be homogeneous or can contain specific binding sites for target ions. The Ion-Imprinting Polymerisation technique is used to develop the selective material which is explained in the next sections.

6.2 Ion Imprinting Polymerisation

Ion imprinting has become a fast-growing technology that has gained much attention recently, especially in the area of materials science. One of them is called ion-imprinted polymers (IIPs), synthesised on the principles of enzyme phenomena whereby a polymer is altered by polymerisation, which takes place in the presence of a template that could later be removed to create cavities to recognise only the analyte of interest [1]. This specific and selective affinity for the target species decreases the chances of competition with other different ions. Ion-imprinted polymers (IIPs) can be defined differently depending on the methods used to synthesise or fabricate them, the functions they play, or the materials used for their synthesis.

Md E. E. Alahi and S. C. Mukhopadhyay, *Smart Nitrate Sensor*, Smart Sensors, Measurement and Instrumentation 35, https://doi.org/10.1007/978-3-030-20095-4_6

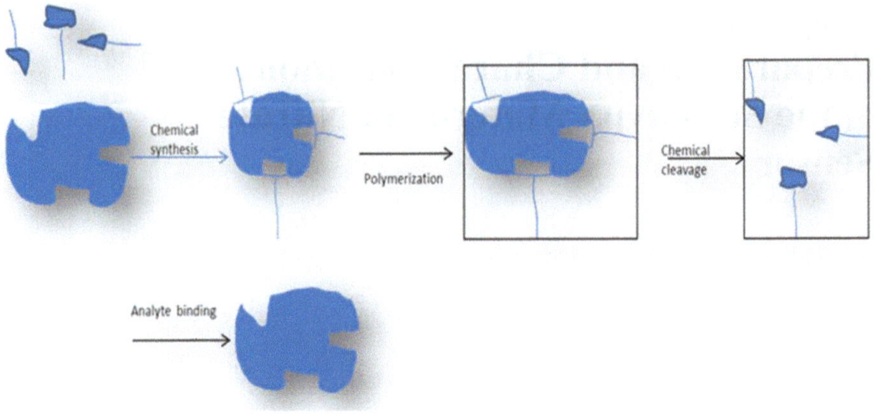

Fig. 6.1 Covalent interaction of IIP

6.2.1 Types of Imprinting Process

The type of interactions involved in the imprinting technique defines the type of imprinted polymer generated and hence the method of imprinting employed [2]. The interaction is discussed in the following subsection.

6.2.1.1 Covalent Interaction

In this type of interaction, the template and the monomer are linked by unbreakable covalent bonds, though such bonds are limited in molecular imprinting [2, 3]. However, even though it provides a stable interaction between the monomer and template, it is highly inefficient as far as rebinding and reuse of the polymers to be synthesised are concerned. Figure 6.1 shows the processes involved in the covalent imprinting process.

6.2.1.2 Non-covalent Interaction

To achieve better rebinding kinetics and rapid template and monomer equilibrium, Mosbach [4] studied another generation of polymers based on non-covalent imprinting. In this method of imprinting, the template and monomer complex is formed by self-assembly. The monomer and template should be complementary, that is, the binding or completion is based on the union of these two by several mechanisms such as hydrogen bonding, hydrophobic interactions, salt bridges, ionic interactions, and van der Waals forces [2, 5]. Figure 6.2 is a diagrammatic representation of non-covalent imprinting.

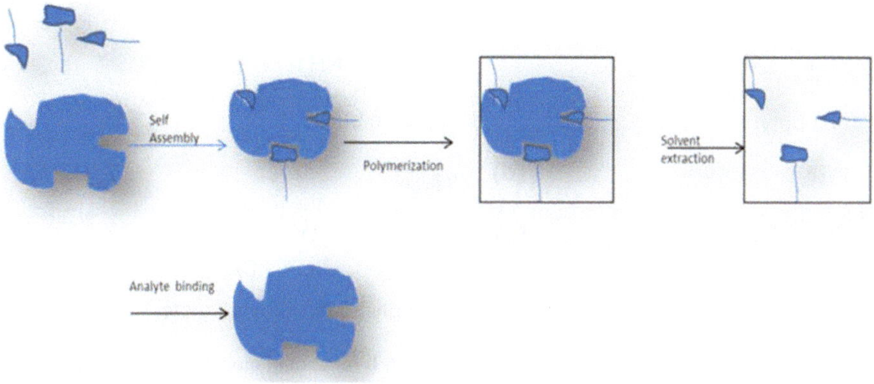

Fig. 6.2 Non-covalent interaction technique

6.2.2 Monomers, Cross-Linkers, Solvents and Initiator for the Imprinting Procedure

6.2.2.1 Monomers

The proper selection of monomer, cross-linker, and solvent affects the performance of the IIPs in their respective applications [6]. The performance is regarding shelf life, utility, rebinding abilities, and reuse [7]. In non-covalent imprinting, the durability of the polymer depends on the functional monomer and template interaction. The monomer should have functional groups that will work towards strengthening the interaction between itself and the template and also the polymer matrix [8].

6.2.2.2 Cross-Linkers

Cross-linkers play a vital role in the imprinting process and create an environment whereby the orientation of the functional monomers and template favours formation of a rigid network where the template and, hence, the cavities are affixed [9]. The more stable these cavities are, the more useful the polymer will be. The cross-linkers should be in higher proportions in the final polymer, and highly branched, or high-molecular-weight crosslinkers have a tendency of producing such stable cavities [10].

6.2.2.3 Solvents

In most cases, solvents are used in the synthesis of IIPs. These include water, toluene, chloroform, acetonitrile, and many more [11]. A suitable solvent dissolves all the

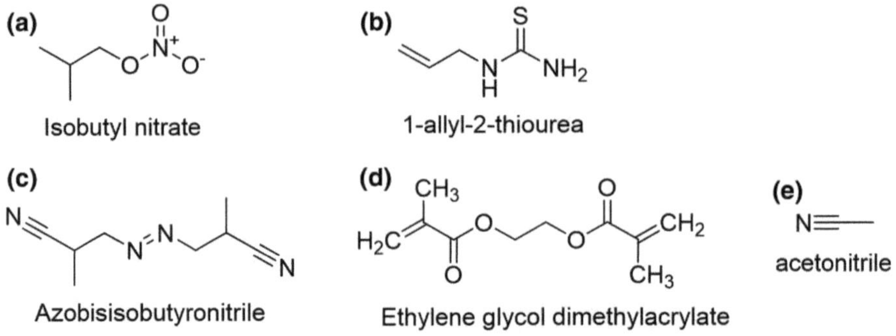

Fig. 6.3 Chemical structure of: **a** Template molecule, **b** functional monomer, **c** initiator, **d** cross-linker and **e** solvent

components of the polymerisation mixture and by extension contributes to the physical makeup of the polymer. The usability of solvents may depend on the hydrogen-bond parameter, dielectric constant, polarity, and solubility, which may increase or decrease the durability of that polymer [12].

6.2.2.4 Initiator

The initiator is used to initiate the polymerisation process. It is a source of chemical species that reacts with a monomer to form a compound which can link with a large number of other monomers during a polymerisation process. Upon the exposure to heat or light, it generates free radicals or initiators. Figure 6.3 represents the chemical structure of various polymerisation components which are used in this research.

6.3 Polymerisation Methods

There are four different polymerisation methods generally used in the imprinting technique that have adequately been studied. Each has got some more attractive features than the others. They are discussed in the subsequent sections.

6.3.1 Bulk Polymerisation

Bulk polymerisation has been preferred in recent times by most researchers [13]. They have limitations such as the non-uniform distribution of particle size, thermal instability, wasting of the polymer at the grinding stage, and destruction of binding sites. They are easy to prepare. The synthesis of IIPs via the bulk polymerisation process includes mixing all the materials: template, monomer, cross-linker, and initiator in a reaction vessel; sometimes the polymerisation process is initiated by UV irradiation.

6.3.2 Precipitation Polymerisation

This type of polymerisation emerges as an excellent alternative to bulk polymerisation, having a high yield and forming particles that are rigid and cross-linked, which helps them not to be readily coagulated; hence, it needs no stabilisers and, therefore, shows specific binding properties because the particles are free of stabiliser molecules. This method gives particles at the 0.3–10 μm range [14], and these highly cross-linked particles are formed with an excess amount of solvent.

6.3.3 Suspension Polymerisation

Generally, this is a complex process whereby the initiator is soluble in the monomer, and these two are insoluble in the continuous phase, which in most cases is water [15]. This method produces micro-beads with a size distribution in the range of 0.5–10 μm, which are enhanced by the rate of stirring. Droplets of the monomer and initiator are suspended in the continuous phase, in the presence of a stabiliser, then polymerised. It is fast and reliable, and polymerisation of particles may be achieved in the space of two hours.

6.3.4 Surface (Emulsion) Polymerisation

The surface polymerisation is an excellent alternative to the other types of polymerisation and one of the highly used polymerisation methods. It annihilates the template and offers very rapid adsorption and desorption kinetics. Surface polymerisation also possesses good stability in acidic conditions which is suitable for metal removal, as most metals are soluble in acidic conditions, which may imply that better removals may take place in those conditions. This method is also able to produce polymers of high molecular weight which helps in the stability of the polymer and is also easy to control and to remove heat.

Table 6.1 Synthesis recipe of IIP and NIP polymerisation

Polymer	Template (mmol)	Functional Monomer (mmol)	EGDMA (mmol)	Acetonitrile (mL)
IIP	1.0	4.0	20.0	4.0
NIP	none	4.0	20.0	4.0

6.4 Materials and Methods

6.4.1 *Chemicals*

Isobutyl Nitrate (IBN) ($C_4H_9NO_3$), 1-Allyl-2-thiourea (AT) ($C_4H_8N2_S$), Ethylene glycol dimethyl acrylate (EGDMA) ($C_{10}H_{14}O_4$), acetonitrile (ACN) (C_2H_3N) and 2, 2′-azobisisobutyronitrile (AIBN) ($C_8H_{12}N_4$) were purchased from Sigma-Aldrich, Australia. Liquid acrylic resin has been purchased from Briture Co. Ltd from Anhui, China. EGDMA and ACN were also purchased from Sigma-Aldrich, Australia as commercial reagents of the highest grade and distillation was not required before use. Deionised (DI) water was collected from a Millipore (18 MΩ cm) water system. A standard nitrate-N sample with a concentration of 100 (mg/L) was also purchased from Sigma-Aldrich. A NitraVer X Nitrogen-Nitrate Reagent Set was purchased from HACH, Australia.

6.4.2 *Apparatus*

A Hach DR/4000 Spectrophotometer was used to validate the concentrations generated using nitrate-N detection [16]. A Hioki IM 3536 LCR meter [17] was used to measure the impedance behaviour of the interdigital sensor. A Mettler Toledo Pro-ISM-IP67 pH meter was used to measure the pH level in the sample water.

6.4.3 *Synthesis of Nitrate Imprinted Polymer Coating*

A conventional synthesis protocol for preparing the ion imprinted polymer, involved dissolving the following chemicals in acetonitrile (4.0 mL): Isobutyl nitrate (IBN) (1 mmol), 1-allyl-2-thiourea (AT) (4.0 mmol), Ethylene glycol dimethyl acrylate (EGDMA) (20 mmol) and Azobisisobutyronitrile (AIBN) (0.12 mmol). IBN and AT were used as the template molecule and functional monomer respectively, for preparing the nitrate derivative-imprinted polymers [18]. The reaction mixture was purged with nitrogen gas for 10 min to remove oxygen. The mixture was polymerised

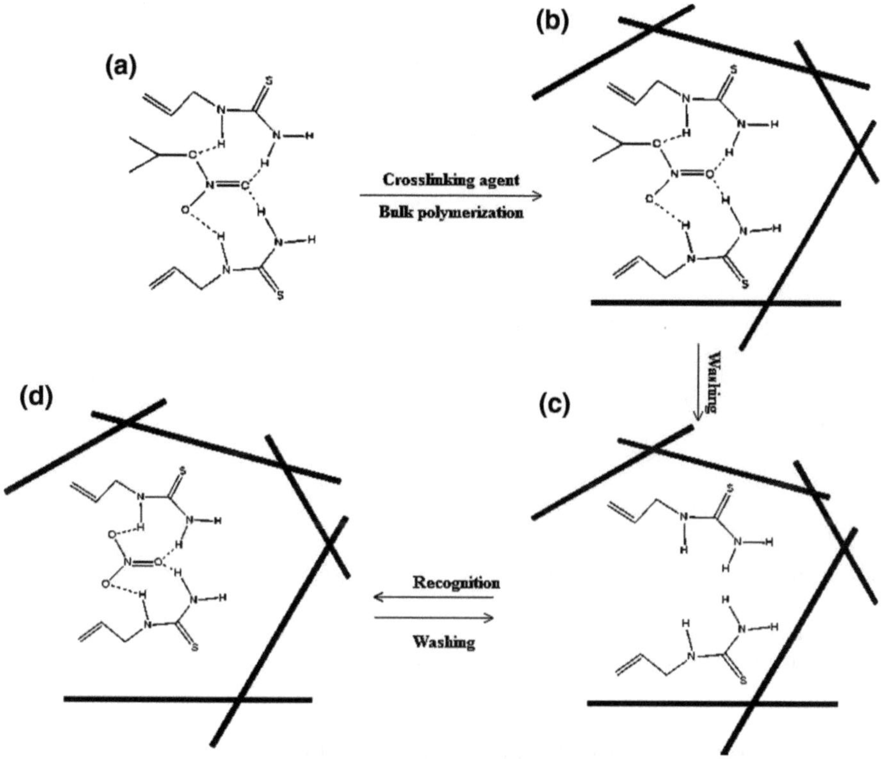

Fig. 6.4 Graphic illustration of IIP with AT as a functional monomer. **a** IBN (template) and AT were mixed insolvent; **b** polymerisation with adding of EGDMA and black lines are indicating the formation of polymer with target ions and functional monomer. EGDMA helps to form the polymer. **c** Creating the cavity for template molecule; **d** binding of nitrate ion. Table 6.1 is given the synthesis recipe of IIP and NIP polymerisation

in a water bath at 55 °C for 16 h, followed by 80 °C for 3 h, under a nitrogen environment. The polymer was washed with methanol to remove any unwanted particles. A mechanical mortar was used for grounding the polymer and filtered to produce a particle size of 75 μm. The template molecule was removed from the polymer particle by washing with a mixture of 500 mL of methanol/triethylamine (4:1, v/v) solvent, followed by 500 mL of methanol for 24 h by Soxhlet extraction equipment. After removal of the template molecule, the deionised water was used to wash the polymer. Finally, it was desiccated in a nitrogen atmosphere at room temperature, scaled to measure the weight and kept in a closed container at 4 °C. The same synthesis protocol was used without the template molecule to develop a non-imprinted polymer (NIP) (Table 6.1). Figure 6.4 illustrates the polymer synthesis process.

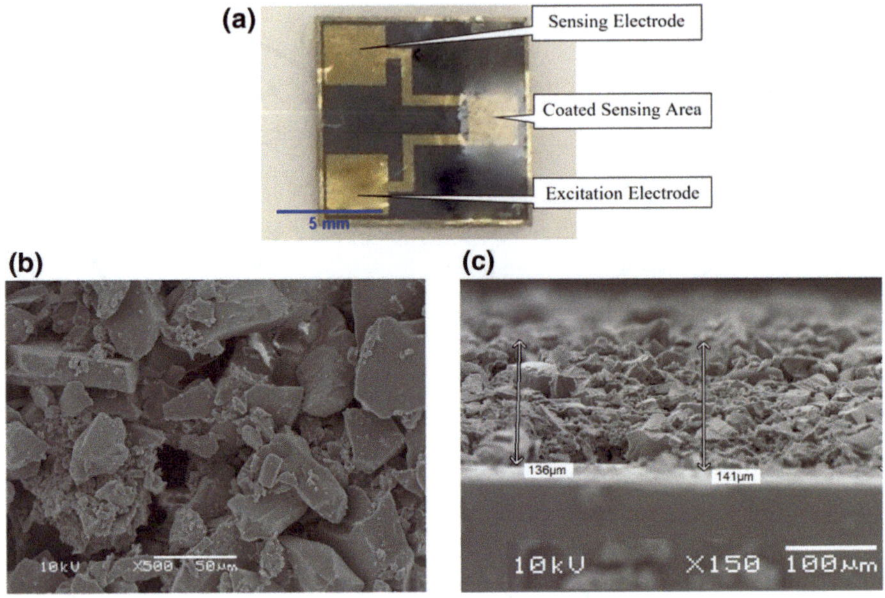

Fig. 6.5 **a** IIP coating on the sensing surface, **b** top scanning electron microscope (SEM) view of the the coating surface, **c** height of the coating as a SEM image

6.4.4 Functionalizing the Polymer Coating

Acetone was used to remove any unwanted material on the sensing surface. 50 µL of acrylic resin, 250 µL of acetone and 400 mg of IIP polymer were mixed to prepare a suspension. PTL-MM01 Dip Coater was used to coat on the sensing surface. The sensor was dipped in the suspension and was withdrawn at a rate of 200 mm/min to achieve a uniform coating on the sensing surface. Due to the evaporation of the acetone, the suspension became the solid coating on the sensing surface. Figure 6.2a shows the IIP polymer coating on the sensing area. The scanning electron microscope (SEM) images of Fig. 6.5b, c are indicating the coating on the sensing area. The height of the coating surface is <150 µm, which is sufficient to measure the impedimetric behaviour of the sensor. The estimated cost of the imprinted polymer, which was used on the coating surface, is <USD 1.

6.4.5 Sorption Study

This study investigates the affinity of nitrate-N ions and measures the adsorption capacity with IIP and NIP. Ten mL of 1, 2.5, 5, 7.5, 10, 12.5, 15, 17.5 and 20 mg/L standard samples of nitrate-N were placed into nine different flasks. Ten mg of dried

IIP polymer was added to the sample solutions. The mixed solution was incubated by stirring in a shaker at 22 °C for 1 h. After incubation, the supernatant was sampled. The whole process was repeated with NIP polymer by using similar standard samples. A 0.22 μm filter was used to filter out the samples. The final filtered solution was analysed. The number of ions bound to the IIP was calculated using the following equation:

$$Q = \frac{(Ci - Cf) \times V}{m} \tag{3}$$

where, Q(mg g^{-1}) = Mass of nitrate adsorbed per gram of IIP, C_i(mg/L) = Initial concentration of nitrate-N; C_f(mg/L) = Final concentration of nitrate-N after adsorption. V(L) = Volume of the adsorption mixture, and m(g) = Mass of the polymer used.

6.4.6 *Static Sorption Time*

This study measures the binding kinetics of nitrate-N IIP and the time it is required to complete the bindings. A 10 ml sample of 10 mg/L nitrate-N was added into six separate 50 mL glass flasks containing 10 mg of dried IIP. All the samples were stirred in a shaker (150 rpm) for 1,3,5,7,9,11,13 and 15 min at a similar temperature and humidity. Consequently, all the mixtures were centrifuged and the supernatant was analysed.

6.4.7 *Selectivity Test*

Fifty mg of IIP polymer was mixed with 50 ml of a solution of sodium nitrite ($NaNO_2$), potassium nitrate (KNO_3) sodium sulfate (Na_2SO_4) and DI water, where the concentration of nitrate-N, nitrite-N and sulfate was 20 mg/L respectively. The mixture was incubated for 1 h with continuous stirring at room temperature. The samples were centrifuged after the incubation, and the supernatant was collected to analyse further for selectivity test. Dionex 2000 Ion Chromatography was used to conduct the analysis.

6.4.8 *A Binding Procedure of Coated Sensor*

The polymer has imprinted cavities which adsorb the nitrate-N ions on the sensing surface from the aqueous medium. Figure 6.6 illustrates the adsorption process of nitrate ions. Ten μl of sample water was pipetted on the coated sensing area. Due to

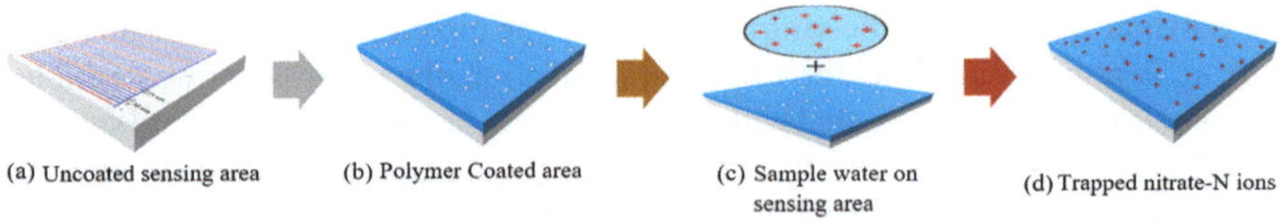

Fig. 6.6 Diagram describing the adsorption process of nitrate-N ions **a** uncoated sensing surface, **b** coated sensing surface, **c** sample water is adding on the sensing surface, **d** ions are trapped on the sensing surface. White markers are indicating the imprinted cavities on the polymer, and red markers are indicating the nitrate-N ions

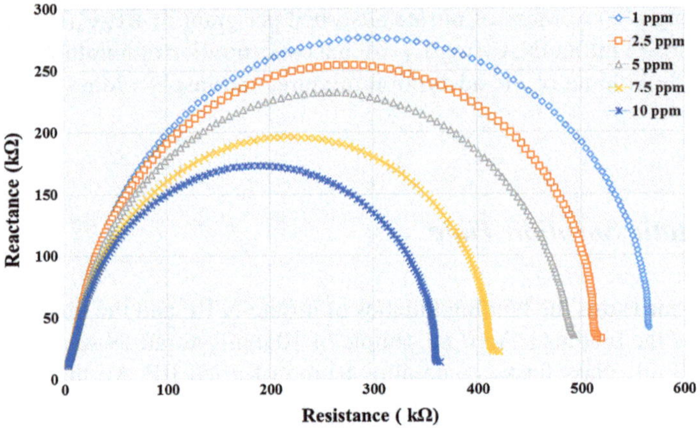

Fig. 6.7 Nyquist plot for different concentration

the hydrogen bond of the imprinted polymer, the nitrate-N ions were entrapped by the cavities. The sensor took seven minutes to absorb the maximum number of ions and took 30 min to dry the wet surface. After drying the coating surface, the EIS technique was applied to measure the changing behaviour of the impedance data.

6.4.9 EIS Measurement of the Coated Sensor

Samples containing 1, 2.5, 5, 7.5 and 10 mg/L of nitrate-N were analysed, and the corresponding impedance was measured by an LCR meter. A Cole-Cole plot or Nyquist plot was created for impedance measurement, which has shown in Fig. 6.7. It is seen from Fig. 6.7 that the radius of the plot is decreasing due to the increase in solution concentration. Figure 6.8 shows the change of reactance on frequency. There is a significant difference in reactance between all the concentrations in the frequency range 40–220 Hz. The maximum difference of reactance was used to create the calibration curve, to measure the unknown sample for a specific frequency.

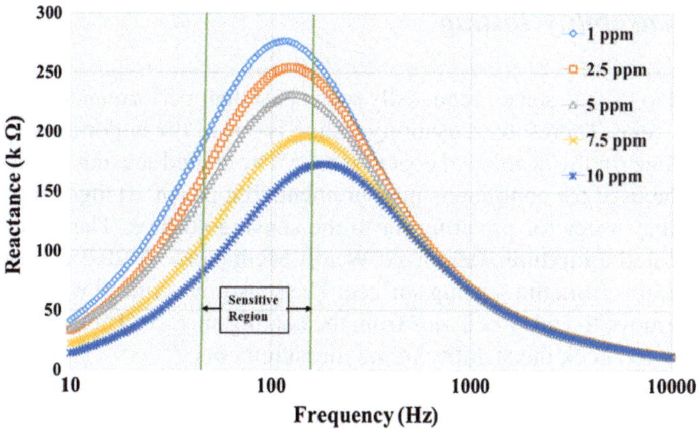

Fig. 6.8 Reactance versus Frequency for various concentrations

The coated sensor was exposed in the water, and according to the imprinting polymer principle, the nitrate-N ions were trapped by the recognition sites. The sensor was exposed for 7 min to finish the entrapping. During the EIS measurement, the electrical intensity bulges through that imprinted polymer, and corresponding impedance changed. The reactance or imaginary impedance are considered due to the significant change for different concentration. The number of ions is different for different nitrate-N concentration, which leads to different reactance measured by the LCR meter.

6.4.10 Unknown Sample Measurement

This experiment was conducted with different water samples, which were collected from various sampling locations in Sydney, Australia using the developed coated sensor. The natural nitrate concentration was not high enough to do the experiments for high concentrations. Therefore, additional nitrate was added to these samples to elevate the nitrate-N concentration to the desired concentration range, without changing the other parameters. An exception was sample 5, which did not have any nitrate-N added. The spectrometric method was used to validate the sensor measurements. The measurement was undertaken five times, and the results averaged for analysis.

6.4.11 Reusability Testing

It is required to use the sensor repeatedly with consistent performance to monitor the water in real-time. Therefore, reusability testing is one of the important experiments, which was done for the developed coated sensor. The coated sensing surface requires cleaning to be used for continuous measurement. Ten μL of 10 mg/L nitrate-N was used as sample water for pipetting on to the sensing surface. The ratio of Acetic Acid has created a mixture: Deionized Water: Methanol = 1:20:2 (v/v) to remove the adsorbed ions from the sensing surface. The sensor was rinsed with the prepared mixture to remove the adsorbed ions from the coating surface. The process recorded for five times to check the stability of the measurements.

6.4.12 Non-linear Least Square Curve Fitting

An EIS Spectrum Analyser was used to analyse the impedance spectra. The algorithm of electrochemical spectrum analyzer was used for theoretical calculations to estimate the equivalent circuit parameters. The measured impedance from the sample solution and theoretically estimated values were matched by a complex algorithm, which is called the nonlinear least square fitting technique. That complex algorithm also estimated the equivalent circuit parameters. The statistical approach used to calculate the residual mean square r_c^2, used the following relationship:

$$r_c^2 = \sum_{i=1}^{N} \frac{\left(Z_i' - Z_{i_{calc}}'\right)^2 + \left(Z_i'' - Z_{i_{calc}}''\right)^2}{Z_i'^2 + Z_{i_{calc}}''^2} \tag{4}$$

where $Z_i^{'}$ and $Z_i^{''}$provide observed real impedance and imaginary impedance respectively, $Z_{icalc}^{'}$ and $Z_{icalc}^{'}$provide calculated real impedance and imaginary impedance respectively, r_c^2 explains the deviation in between the experimentally observed data and optimal solution. The proposed equivalent circuit with the corresponding mathematical model is used for a number of times to optimise the solution to get the best-fitted curve. The *residual mean square* is the error rate which estimates how good the fitting model is with the equivalent circuit parameters and calculating those parameters with the error rate <5%.

6.4.13 Comparison of Coated Sensor and ISE

Our earlier work [4] reported a smart sensing system which measured the impedance of an interdigital sensor. The smart sensing system is used to calculate the reactance or imaginary impedance of the coated sensor. In order to calculate the imaginary impedance of the sensor, the microcontroller calculates the phase difference of the input voltage and current. The calculated phase difference is used to calculate the reactance of the coated sensor. The detailed of the sensing system has been discussed in our earlier work [19]. The sensing system has been used to measure the concentration of nitrate-N in real water samples and compared these results with sensION+9662 Nitrate Ion selective electrode [20], which is a commercial sensor. Both results are compared with results from standard laboratory measurement to calculate an error rate. Compared to our earlier work [4]; Arduino Uno Wi-Fi embeds the current sensing system. One of the sample data also sent to IoT based web server to monitor the sensor data in real-time.

6.5 Results and Discussions

6.5.1 Sorption Studies

It is required to investigate the affinity of nitrate-N ions with IIP and NIP. The adsorption studies of nitrate-N ions to the IIP and NIP are shown in Fig. 6.9a. The figure illustrates that the number of nitrate-N ions bound to the IIP increased with increases in the sample nitrate-N concentration up to a concentration of 11 mg/L. Above 11 mg/L, the polymer became saturated, and the curve was parallel to the concentration axis. However, the amount of nitrate-N ions bound to NIP at equilibrium was 7 mg/L and reached saturation. The result showed that the amount of nitrate-N ions bound to IIP was higher than NIP at higher concentrations because of the more specific binding sites of IIPs are generated due to imprinting polymerisation. The countries which follow the acceptable drinking limit of EPA can also accept this saturated limit and the developed polymer material is useful for coating.

6.5.2 Uptake Kinetics Studies

An adsorption kinetics study helps to measure the adsorption time, which is required to trap all the target ions on the developed IIP. Figure 6.9b shows the plot for the uptake kinetic study of nitrate-N to IIP. The concentrations were chosen according to the experiment, which was undertaken in the previous section. The results indicate that the nitrate-N IIP maintains fast trapped uptake kinetics, and the equilibrium of the binding was reached in almost 7 min. The time also counts as the response time of the coated sensor, due to the use of the polymer as a coating on the sensing surface. The EIS measurement of the sensor is less than 5 s.

6.5.3 Study of Selectivity Test

Figure 6.10 illustrates that the peak area of the nitrate-N ions (retention time of 8.5 min) was distinctively reduced, compared to other ions (sulphate and nitrite). The retention time for nitrite and sulphate ions are at 5.39 and 6.58 min. The imprinted polymer has a high affinity for nitrate in aqueous media (pH 6.60). The binding of other analyte anions (nitrite and sulphate) have a relatively low affinity. Nitrite and sulphate are the other interfering ions which have a similar ionic structure to nitrate-N ions. With this in mind, a subsequent experiment was undertaken where 11 mg/L of nitrate-N, nitrite-N, and sulphate ions were placed into 3 different containers. From Sect. 3.1, it is known that at nitrate-N concentrations of up to 11 mg/L the majority of the nitrate-N can be adsorbed by the IIP. Therefore, a similar amount of nitrite-N and sulphate ions are considered for the experiment. It is seen from Table 6.2 that the IIP bound 90% of ions in the presence of nitrite and sulphate ions. In the absence of these additional ions, the IIP adsorbs 98% of the nitrate-N ions. The result suggests that the nitrate-N has a higher affinity towards the developed imprinted polymer compare to similarly structured ions. IBN has one functional group which was precisely similar as nitrate-N and removal of IBN from the imprinted polymer created the shape of nitrate-N which has a higher affinity of nitrate-N only. Nitrite-N and Sulphate ions

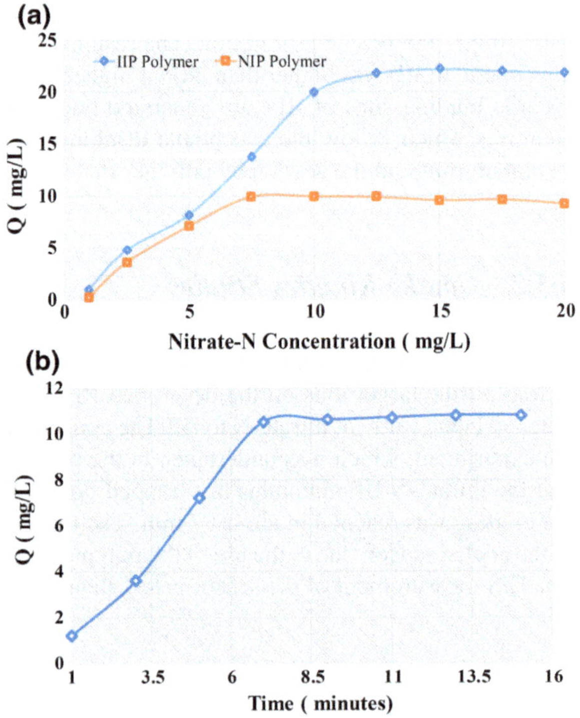

Fig. 6.9 **a** Sorption study for IIP and NIP polymer, and **b** uptake kinetics study

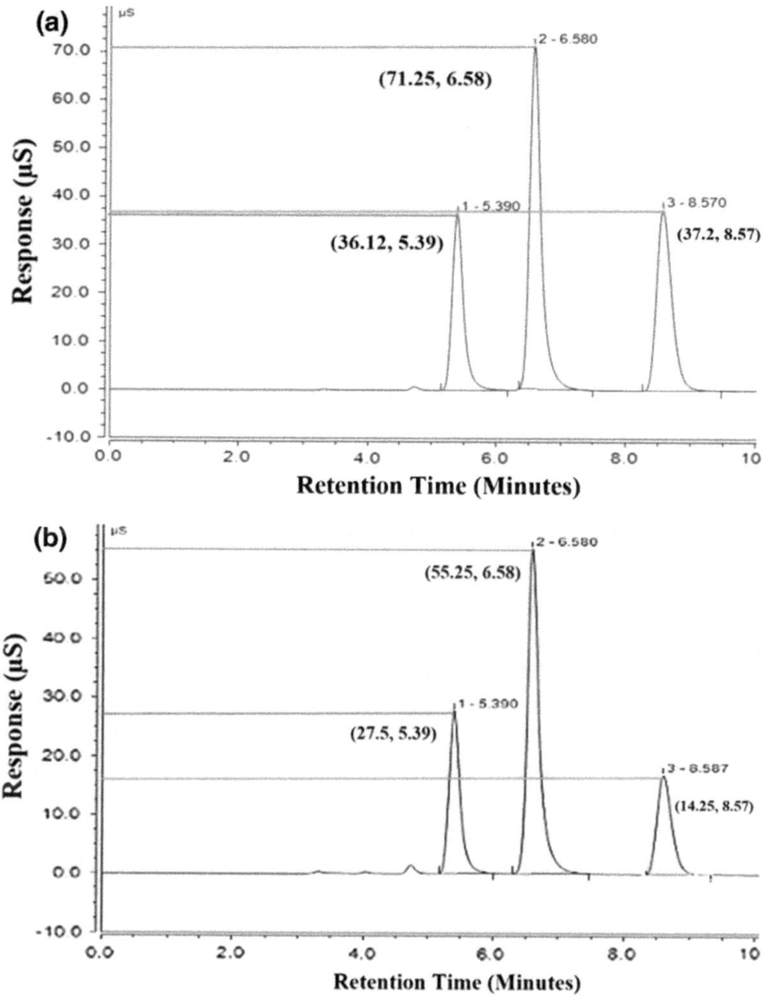

Fig. 6.10 Chromatograms of sample mixture in the imprinted polymer, **a** mixed solution before incubation, **b** after incubation

Table 6.2 Binding of nitrate-N in aqueous media

	Nitrate-N (%)	Nitrite-N (%)	Sulphate (%)
Mixture of different ions	90	6	4
Only Nitrate-N	98	–	–

have different shapes and sizes which makes the developed IIP specific for nitrate-N ions.

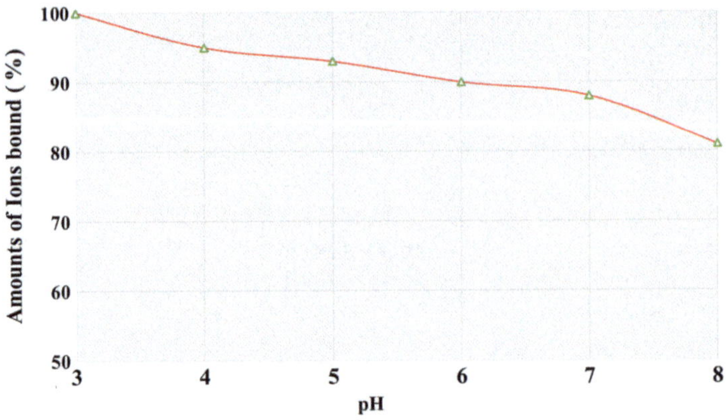

Fig. 6.11 Effect of pH of the measurement

6.5.4 pH-Dependent Binding Profile

The performance of the IIP coated sensor was studied in different pH solutions to observe the effect of pH. The pH was varied by the standard buffer solution to cover a range of 3.0–8.0. The surrounding ionic strength of the polymer will change significantly when it is exposed to a pH solution. It has been reported [21] that the increase in pH usually increases the swelling of the polymer. This phenomenon can be observed on the developed IIP as well and the results are illustrated in Fig. 6.11. The number of ions bound by the sensor increased as pH decreased. As the pH increased, the number of ions bound reduced significantly. The sensor was operated at a pH value 6.60, where the IIP coated sensor bound nearly 90% of the ions which is acceptable in New Zealand water.

6.5.5 Calibration Standard

It is seen from Fig. 6.12 that at 110 Hz, the difference of reactance change for various concentrations is at maximum. Therefore, reactance for various concentrations is taken at that frequency to develop the calibration curve to measure an unknown sample. The nitrate-N concentration is represented on the x-axis, and corresponding reactance is represented on the y-axis. Figure 6.12 indicates the developed calibration graph which is linear ($r^2 = 0.99$). This calibration standard will be used to measure the nitrate-N concentration of any unknown water sample.

The final formula to calculate the unknown sample-

$$\mathbf{C} = \frac{\mathbf{R} - \mathbf{289418}}{\mathbf{-12811}} \tag{5}$$

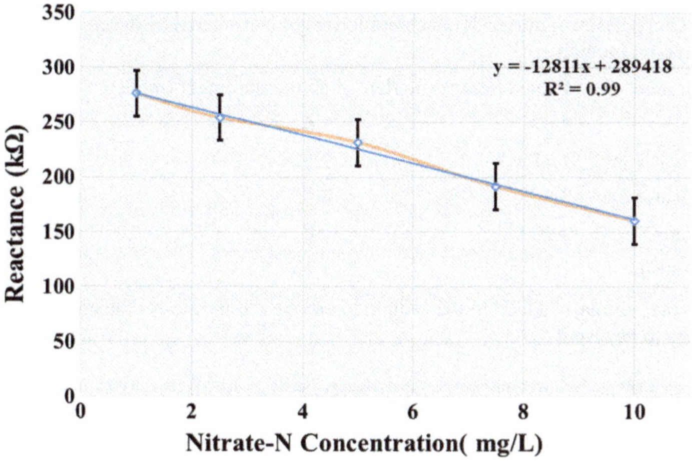

Fig. 6.12 Calibration standard for nitrate-N measurement

where, C is the unknown sample in mg/L, and R is the measured reactance (Ω) from EIS measurement. This standard is applicable when the coated sensor operates at 110 Hz. The LOD and limit of quantification (LOQ) were 1.06 and 3.21 mg/L, respectively. Therefore, 1.06 mg/L is the lowest concentration, which can be measured by the developed coated sensor, and 3.21 mg/L can be measured reliably at the specific constant temperature and humidity.

6.5.6 Unknown Sample Measurement

Equation 5 used to measure an unknown sample by the developed coated sensor. Sample 5 was taken as one of the unknown samples, and the coated sensor was used to measure the concentration. It was found that the reactance of the sample, R = 275,325.9 Ω. The measured R was used in Eq. 5 and the final concentration was C = 1.1 mg/L. Table 6.3 shows that the coated sensor accurately measured the sample concentration across a wide range of nitrate-N concentrations, with an error rate of less than 1%, when the concentration was more than 3 mg/L. This supports the fact that the LOQ of the sensor is 3.21 mg/L. When the concentration is on the lower side (1–3 mg/L), the error rates are closer to 5%, which is still acceptable for this measuring purpose. In New Zealand and Australia, the concentration of nitrate-N in surface water covers this range, and the coated sensor can measure this range with a reasonable amount of error, which will be acceptable statistically.

Table 6.3 Comparison of nitrate-N concentration (mg/L) measured by the coated sensor and using the spectrometric method

Sample number	Coated Sensor (mg/L)	Lab measurement (mg/L)	Error rate (%)
1 (Lake water)[a]	3.57	3.6	0.83
2 (Stream water 1)[a]	2.47	2.5	1.19
3 (Stream water 2)[a]	8.5	8.55	0.58
4 (River water)[a]	9.96	10.0	0.4
5 (Canal water)	1.1	1.15	4.34

[a]Had additional nitrate-N added to the sample to elevate the nitrate-N concentration to within the desired range to be tested

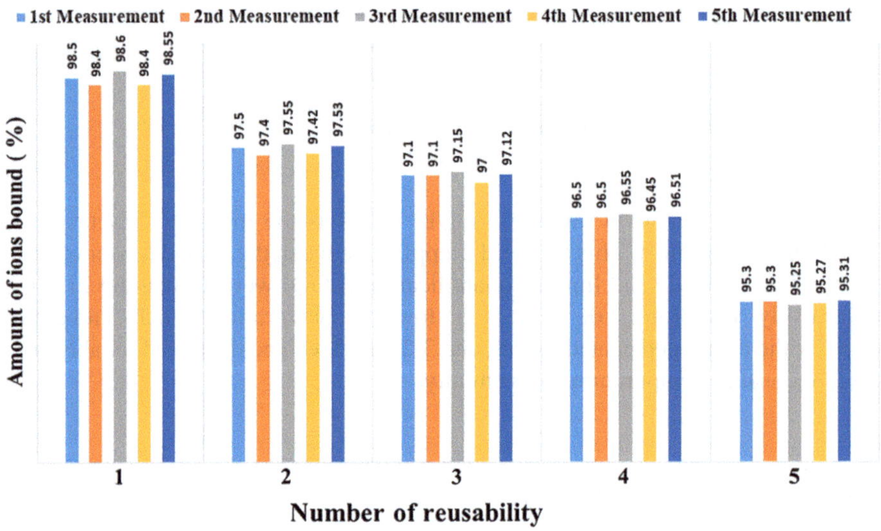

Fig. 6.13 Reusability of the sensor

6.5.7 Reusability of the Sensor

The number of ions bound on the sensing surface was measured and plotted in Fig. 6.13. This showed that the sensor can be reused accurately five times before measurement errors occur. More than 95% of the nitrate-N ions can be rebound by the IIP coating surface for the first five measurements. After five measurements, the sensor recorded errors of >10%. It is likely; some of them the recognition sites were damaged due to the washing of the sensing surface repetitively. Therefore, some of the ions are not binding by the recognition sites, which were binding earlier. It is also seen that the repetitive measurements were also stable.

Table 6.4 Comparison of measurement between the coated sensor and ISE sensor

Sample	Coated sensor + Sensing system		ISE sensor		Lab measurement (mg/L)
	Conc.(mg/L)	Error rate (%)	Couc.(mg/L)	Error rate (%)	
1 (Tap water)	1.22	1.67	1.26	5	1.20
2 (Stream Water 1)	1.67	1.76	1.73	1.76	1.7
3 (Stream Water 2)	2.12	1.39	2.20	2.32	2.15
4 (Lake water)	2.35	2.17	2.34	1.73	2.3
5 (River water)	1.80	2.70	1.75	5	1.85

6.5.8 Comparison of Coated Sensor and ISE

Table 6.4 compares results from the coated sensor combined with the sensing system to the ISE sensor and provides a measurement of error rate (%). The coated sensor combined with the sensing system measured nitrate-N with an error rate of less than 5%. The ISE sensor recorded higher measurement errors compared to the laboratory measurement. Sample 4, which was collected from a lake, was used to transfer through the developed sensing system with the integrated WiFi to an IoT based web server and the real-time data are shown in Fig. 6.14. Therefore, the developed sensor can measure the nitrate-N concentration in the actual sample and transfer the measured data through a low-cost device, which can be used as a part of the distributed network in future.

6.5.9 Non-linear Least Square Curve Fitting

The estimated circuit parameters and the model have been given in Table 6.4. The fitted curve and simulated results are shown in Fig. 6.15a, b. It is seen in Fig. 6.15a that the red markers are experimentally obtained results, and the solid green line represents the fitted curve for the equivalent circuit. It is also seen that the curve nicely fits with the experimentally obtained results. Figure 6.15b represents the fitting of simulated results (blue markers) from the developed model, which is also matched with the measured or experimental results. Figure 6.15a also shows the equivalent circuit parameters which are imposed on the fitted curve for the coated sensor. C1 is the double layer capacitance, which formed on the interface between an electrode and its surrounding electrolyte. This double layer is formed as ions from the solution adsorb onto the electrode surface. The positive and negative electrodes are parted from the

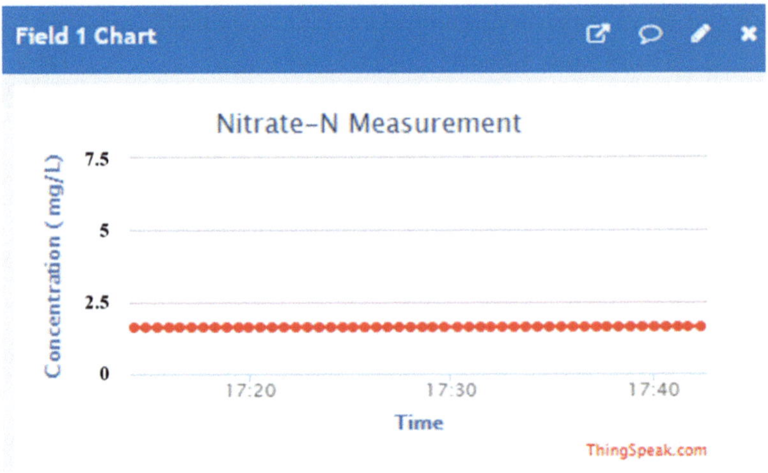

Fig. 6.14 Real time data transferred to an IoT server

Table 6.5 The coated sensor with the equivalent circuit parameters in different concentration

Concentration (mg/L)	C1(nF)	R1(Ω)	R2 (Ω)	P1	n1
1	2.4171	556020	17328	7.51E-09	0.91
2.5	2.4078	508490	14433	6.13E-09	0.93
5	2.4827	472890	14513	1.07E-08	0.88
7.5	2.5135	397430	17904	9.14E-09	0.89
10	2.4366	348340	14544	6.78E-09	0.92

charged ions by an insulating space, often about angstroms. R1 is the charge transfer resistance. In this study, the double layer capacitance and charge transfer resistance are the key electrical parameters in determining impedance change, when analysing system kinetics.

Figure 6.16 is the calibration standard developed from R1 of Table 6.5. One of the equivalent circuit parameters was used to develop this calibration standard which is linear ($r^2 = 0.98$). This calibration standard can be used to measure an unknown sample, which is not frequency dependent. The earlier calibration standard is frequency dependent, whereas is the calibration standard from CNLS analysis was frequency independent. Nonlinear curve fitting method requires large computation memory and time, which forces to use the earlier developed frequency dependent calibration curve with a smart sensing system, to measure nitrate-N concentration from an unknown sample.

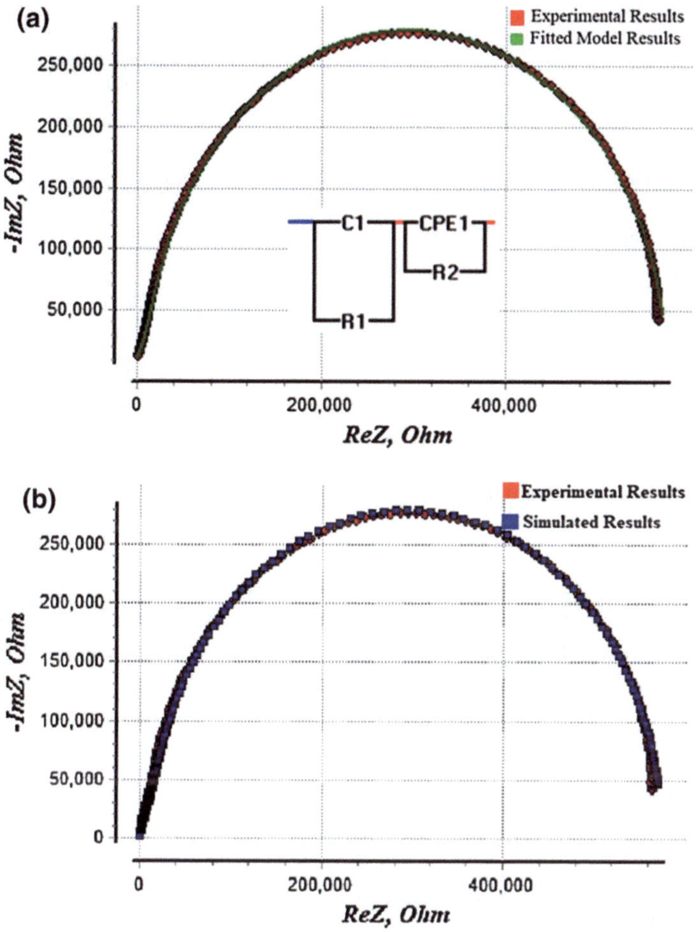

Fig. 6.15 **a** Experimental results and fitted model results are fitted together; **b** Experimental results and simulated results are fitted together from the developed model

6.6 Chapter Summary

Ion-imprinted polymer-based nitrate-N has been developed and used as a coating material on the sensing surface. The interdigital capacitive sensor has been used for detection purposes where the detection limit is 1–10 mg/L. The sensing surface was functionalized by an acrylic resin with an embedded coating material to introduce selectivity for nitrate-N. A calibration curve has been developed to measure an unknown sample. The measurement of unknown water samples has shown good results, with an error rate of <5%. The sensor can be reused up to 5 times before the error rate becomes unacceptable. CNLS curve fitting was also applied to interpret the experimentally obtained impedance spectra. An earlier reported sensing system was

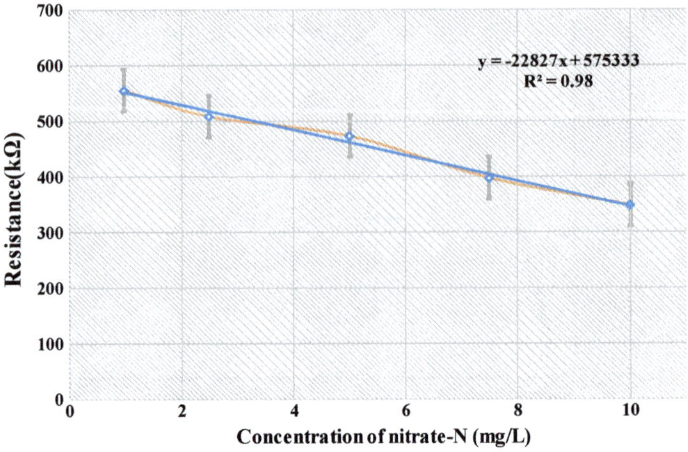

Fig. 6.16 Calibration standard from CNLS analysis

also used to measure the nitrate-N concentration in real time and transfer the data to a web server, which has excellent potential to develop a vast distributed network. Future research will focus on increasing the reusability of the sensor, so that, it will be more useful for real-time water monitoring.

References

1. L.D. Mafu, T.A. Msagati, B.B. Mamba, Ion-imprinted polymers for environmental monitoring of inorganic pollutants: synthesis, characterisation, and applications. Environ. Sci. Pollut. Res. **20**(2), 790 (2013)
2. G. Wulff. The covalent and other stoichiometric approaches. *Molecularly Imprinted Materials: Science and Technology*, ed. by M. Yan, O. Ramström (Marcel Dekker: New York, 2005). pp. 59–92
3. V. Vishnuvardhan, K. Prathish, G. Naidu, T.P. Rao, Fabrication and topographical analysis of non-covalently imprinted polymer inclusion membranes for the selective sensing of pinacolyl methylphosphonate—a simulant of soman. Electrochim. Acta **52**(24), 6922 (2007)
4. K. Mosbach, Y. Yu, J. Andersch, L. Ye, Generation of new enzyme inhibitors using imprinted binding sites: the anti-idiotypic approach, a step toward the next generation of molecular imprinting. J. Am. Chem. Soc. **123**(49), 12420 (2001)
5. O. Ramstroem, L.I. Andersson, K. Mosbach, Recognition sites incorporating both pyridinyl and carboxy functionalities prepared by molecular imprinting. J Organic Chem. **58**(26), 7562 (1993)
6. N. Candan, N. Tüzmen, M. Andac, C.A. Andac, R. Say, A. Denizli, Cadmium removal out of human plasma using ion-imprinted beads in a magnetic column. Mater. Sci. Eng., C **29**(1), 144 (2009)
7. G. Wulff, Enzyme-like catalysis by molecularly imprinted polymers. Chem. Rev. **102**(1), 1 (2002)
8. L. Chen, S. Xu, J. Li, Recent advances in molecular imprinting technology: current status, challenges and highlighted applications. Chem. Soc. Rev. **40**(5), 2922 (2011)

9. E. Caro, R. Marcé, F. Borrull, P. Cormack, D. Sherrington, Application of molecularly imprinted polymers to solid-phase extraction of compounds from environmental and biological samples. TrAC Trends Anal. Chem. **25**(2), 143 (2006)
10. J. Matsui, I.A. Nicholls, T. Takeuchi, K. Mosbach, I. Karube, Metal ion mediated recognition in molecularly imprinted polymers. Anal. Chim. Acta **335**(1–2), 71 (1996)
11. H. Su, J. Li, T. Tan, Adsorption mechanism for imprinted ion (ni2+) of the surface molecular imprinting adsorbent (smia). Biochem. Eng. J. **39**(3), 503 (2008)
12. T. Li, S. Chen, H. Li, Q. Li, L. Wu, Preparation of an ion-imprinted fiber for the selective removal of cu2+. Langmuir **27**(11), 6753 (2011)
13. A.C. Mesquita, M.N. Mori, L.G.A. e Silva. Polymerization of vinyl acetate in bulk and emulsion by gamma irradiation. Rad. Phys. Chem. 71(1–2), 253 (2004)
14. K.-C. Ho, W.-M. Yeh, T.-S. Tung, J.-Y. Liao, Amperometric detection of morphine based on poly (3, 4-ethylenedioxythiophene) immobilized molecularly imprinted polymer particles prepared by precipitation polymerization. Anal. Chim. Acta **542**(1), 90 (2005)
15. P.J. Dowding, B. Vincent, Suspension polymerisation to form polymer beads. Colloids Surf., A **161**(2), 259 (2000)
16. HACH. "NITRATE: Chromotropic Acid Method." https://www.hach.com/asset-get.download.jsa?id=7639984361
17. Hioki. "HIOKI IM3536 LCR Meter." https://www.hioki.com/hdfile.jsp?id=11853. Accessed 28 April 2017
18. A. Kugimiya, H. Takei, Preparation of molecularly imprinted polymers with thiourea group for phosphate. Anal. Chim. Acta **564**(2), 179–183 (2006)
19. M.E.E. Alahi, X. Li, S. Mukhopadhyay, L. Burkitt, A temperature compensated smart nitrate-sensor for agricultural industry, in *IEEE Transactions on industrial electronics,* (2017)
20. Hach. sensION+9662 Nitrate Ion Selective Electrode (ISE)—sensor only. https://au.hach.com/sension-9662-nitrate-ion-selective-electrode-ise-sensor-only/product-accessories?id=14533836405. Accessed 12 July 2017
21. N.W. Turner et al., Effect of the solvent on recognition properties of molecularly imprinted polymer specific for ochratoxin A. Biosens. Bioelectron. **20**(6), 1060–1067 (2004)

Chapter 7
IoT Enabled Smart Sensing System

7.1 Introduction

Monitoring the nitrate concentration in the field is an excellent ability for a water-monitoring study. An Interdigital FR4-based capacitive sensor is characterised by nitrate concentration and reported in this chapter. The concentration range of nitrate is 0–40 ppm (mg/L). Different unknown samples were collected from various sources for measurement, and standard UV-Spectrometry was used for validation. An Internet of Things (IoT) enabled smart sensing node has been proposed to measure the rapid nitrate concentration and transferring the data through the gateway to a user-defined cloud server. It can be installed in any locations to collect water, such as a lake, stream, or river. The system is entirely autonomous and solar powered, robust, and trialled in the field successfully. A simple moving-average smoothing algorithm is used to filtering the collected data in the cloud server side. WiFi protocol and the LoRa protocol are compared regarding power consumption. The developed system is tested in the field continuously, and the result validated with standard UV-Spectrometry. The developed smart system has great potential and can be readily deployable in the sampling locations. It also offers new possibilities for both spatial and temporal analysis for nitrate concentration.

7.2 Materials and Methods

The operating principle of planar interdigital sensor [1] is similar like two parallel plate capacitors, where the bare electrodes open up to one side to provide single-sided access to the material under test (MUT) 1. The generated electric field lines penetrate the MUT, and the impedance of the sensor will be changed. The sensor's impedance becomes a function of the material's properties. Therefore, the properties of the system can be estimated by measuring the impedance of the sensor. The term "interdigital" refers to a digit like or finger-like the periodic pattern of parallel elec-

Md E. E. Alahi and S. C. Mukhopadhyay, *Smart Nitrate Sensor*, Smart Sensors, Measurement and Instrumentation 35, https://doi.org/10.1007/978-3-030-20095-4_7

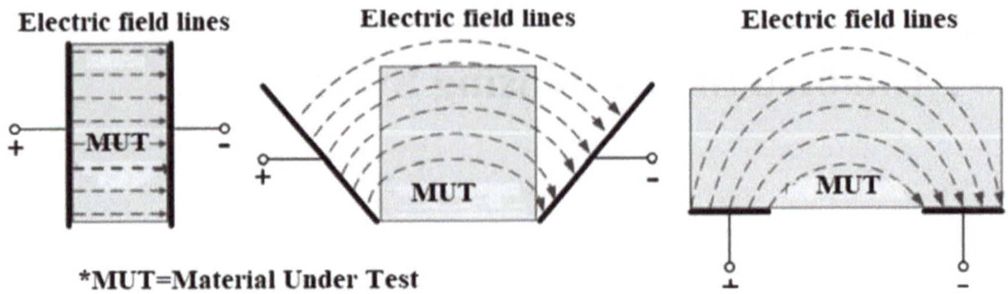

Fig. 7.1 Electric field lines of the parallel plate capacitor and planar interdigital sensor

Fig. 7.2 FR-4 interdigital sensor

trodes that helps to build up the impedance associated with the provided electric field that penetrates the sample material 1. An excitation alternating current (AC) voltage is applied to the positive terminal and negative terminal and electric field forms from the positive terminal to the negative terminal as per Fig. 7.1. The interdigital sensor has been used in a different application for domestic and industrial applications [2–4].

Figure 7.2 shows the FR4 interdigital sensor which is used in the proposed sensing system. The fabrication process is analogous to the process, which is used to follow in developing a printed circuit board (PCB). Copper is used to forming the Electrodes which is a good conductor. The dimension of the sensing area is **33 mm x 17 mm**, which is accessible to dip in the water. The Tin Oxide coating material is used as a layer on the copper electrode to keep the electrodes from corrosion.

7.3 Wireless Sensor Networks (WSNs)

The WSN formed with plenty of small sensing systems, which is named as nodes that collect the information from the environments through various sensors. The system consists of various sensors, microcontroller, and necessary electronics to transmit the collected information to a "gateway" through wireless communication. The gateway can process the collected data and send it to the cloud server where the "Information Management System (IMS)" to analyse the data in real time or for statistical analysis.

- Structure of WSN
  A WSN consists of three different subsystems to build a network. They are called the nodes, the gateway and the Information Management System (IMS).
- Sensor Node
  It is also called sensing system with low cost, size, and low power consumption, sensor, which can be able to get an appropriate measurement from the environment, processing the measured data and sending them directly to the cloud server over the gateway. It consists of the following elements:
  Microcontroller Based System: This is the core of the sensing node, which will be low cost, size, low-power consumption, and low-size chip. Unfortunately, it is challenging to get all these features due to certain constraints, especially if computer power and memory are considered.
  Power Supply Unit: Sensor nodes require a self-directed operational system, which can provide continuous power to run the system all day around. Sensor nodes always are contingent heavily on the power supply unit or power management unit.
  Wireless Communication Network: The WSN network will permit the required communication between the gateway and sensor nodes. They are made of specific standard protocols which have different coverage regions, power consumptions and suitable for different applications.
  Sensors: The sensor node also consists of sensors that convert the external physical parameter from surroundings into an electrical signal to allow the microcontroller to process the filter out data and sending the data during transmission [5].
- The Gateway
  The gateway is also called the "base station" of a WSN network. It is the core of the network, collects data from the smart sensor node, processes and helps to forward to store the data to a cloud server. It also has a computing system, which is based on high power microcontroller or high computing ability. It had to be static and plugged into the main power supply, as it requires more energy than the sensor nodes. It also should have the ability of a wireless communication system that is utilised through the sensing nodes. The gateway guides the collected data to the cloud server through Ethernet, WiFi, and 3G/4G, etc.
- Cloud Server
  A cloud server is the last component of a WSN network. It contains a database, suitable management software for implementing cloud computing and visualise

the processed data for end users. It might be located in a remote computer or any other gateway. Convenient users can get access to the management software through the internet.

- LoRaWAN Protocol for IoT
LoRaWAN (Long range, low power Wireless Area Network) is a data-link layer with extended range, low power, and low bit rate, which is a promising solution for IoT application. LoRa enabled sensor node consumes low energy and transmit few bytes, which is an admirable candidate for being used in many different applications, such as smart cities, smart health care, industry, environmental monitoring, etc. There are two distinct layers: (i) a physical layer, based on radio modulation and also called CSS (Chirp Spread Spectrum); and (ii) a MAC layer protocol which is responsible for getting the access in LoRa architecture [5]. LoRa modulation has the same characteristics of FSK (Frequency Shifting Keying) regarding communication range in between gateway and sensor node. Thus, LoRaWAN is considered as the network architecture, and communication protocol although LoRa supports the long-range link. Battery life, security, the network capacity, the Quality of Service, and reliability of the node are determined by the architecture of the network and the defined protocol [6].

- Energy Harvesting
Sensor nodes can be rarely connected to a fixed power supply; instead, they are the independent, autonomous system with energy harvesting capability. Their consumption of energy must be limited which can be achieved by low-consumption operating modes. Therefore, rechargeable batteries are always used in WSN application with the smart mechanism. For some applications, it is also essential to have a facility to collect the required energy from the environment. These are called energy harvesting technique [7]. Energy sources should be clean and environment-friendly. There are different energy sources are available, such as fluid flows, vibrations, electromagnetic fields, and so on. However, the most used energy source for WSN application is photovoltaic panels or solar panels. The solar panels convert the light source (sunlight or artificial light) into electricity. The battery can store the converted electricity and utilise the energy when there is no sun, such as cloudy day or night. Sensor nodes and smart systems should run cleverly to prolong the battery life and therefore, energy harvesting is an essential factor in WSN application.

7.4 Experimental Setup

- Electrochemical Impedance Spectroscopy (EIS)
Electrochemical Impedance Spectroscopy (EIS) is a highly sensitive method for impedance measurement. There are various methods available to measure the impedance, but Frequency Response Analyzer (FRA) is considered the de facto standard for EIS measurement. FRA requires a small AC voltage with the ampli-

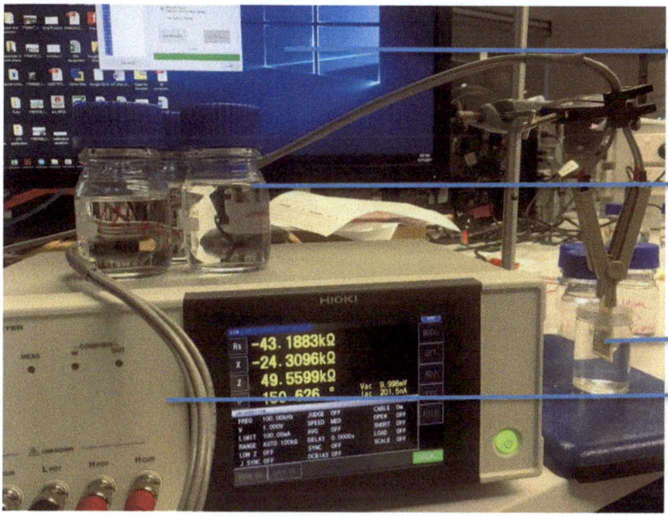

Fig. 7.3 EIS measurement in laboratory conditions

tude of 5–15 mV and frequency of the signal sweeps in a certain range on a direct current (DC) bias voltage. The signal provided on the positive electrode or the working electrode and changed voltage is taken from the negative electrode or sensing electrode. Due to the different characteristics of the different materials, their impedance profile is different, and the following equation presents the impedance:

$$\boldsymbol{Z} = \boldsymbol{R} + j\boldsymbol{X} \tag{7.1}$$

where **Z** is the total impedance (Ω), R is the real part of the impedance (Ω), and X is the imaginary part of the impedance (Ω). The profile of impedance can be characterised graphically which is called a Bode plot or Cole-Cole plot. The FR4 interdigital sensor was characterised by Hioki IM 3536 LCR meter where the frequency swept from 10 Hz to 100 kHz. Standard laboratory temperature and humidity were maintained throughout these experiments. 1, 10, 20, 30, 40 ppm nitrate solution was taken as a standard solution. Deionised water was measured as a control solution. The average pH of the solution was 6.71, which is also preserved in the creek water. Figure 7.3 shows the laboratory setup for the acquisition of data during EIS measurement. Initially, the sensor was characterised to get the impedance profile to develop the calibration standard for nitrate measurement. All the experiments were repetitive for five times to observe the impedance behaviour, and average results were calculated.

- Description of the System
  Figure 7.4 shows the smart sensing system which is proposed to carry out the in-line nitrate analysis. Figure 7.5 illustrates the circuit diagram of the developed system.

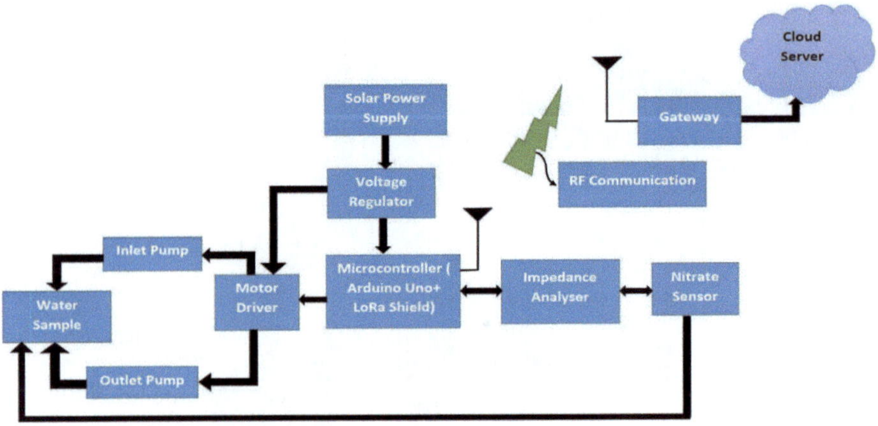

Fig. 7.4 Block diagram of the data transmission

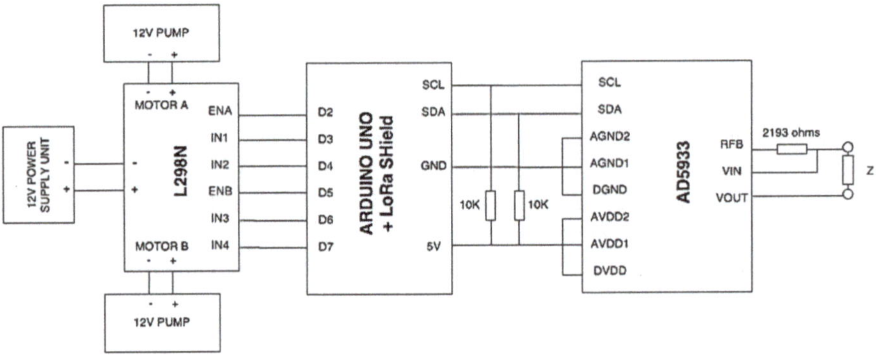

Fig. 7.5 Circuit diagram of the proposed sensing system

Arduino Uno and Arduino Uno Wi-Fi [8] was used as the main microcontroller. An AD5933 [9] is an impedance analyser which has been used to get the impedance data from the sensor. It gets the information in terms of impedance and the phase shift of the sensor. It also can sweep the frequency and gets the corresponding impedance profile, but it is not required in the current application. The operating frequency of the sensor was fixed, and each measurement is taken five times.

A Dragino LoRa shield is required as a remote transceiver to transmit the data to the gateway or receiving the data from the gateway. It allows reaching and sending data to an extremely long-range with low data rates. It is based on the RFM95W/RFM98W [10] and used for 915 MHz transmission/reception. An L298N [11] is a motor drive controller to control the inlet pumps and outlet pumps of the system. The inlet pump is used to bring the water into the reservoir, and the outlet pump is used to empty the reservoir after the measurement.

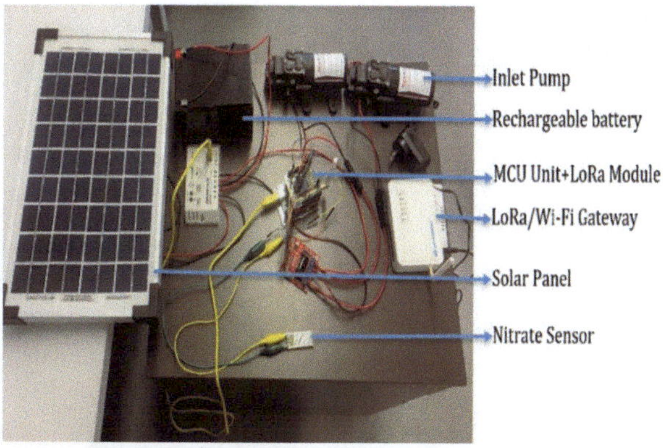

Fig. 7.6 Different parts of the proposed system

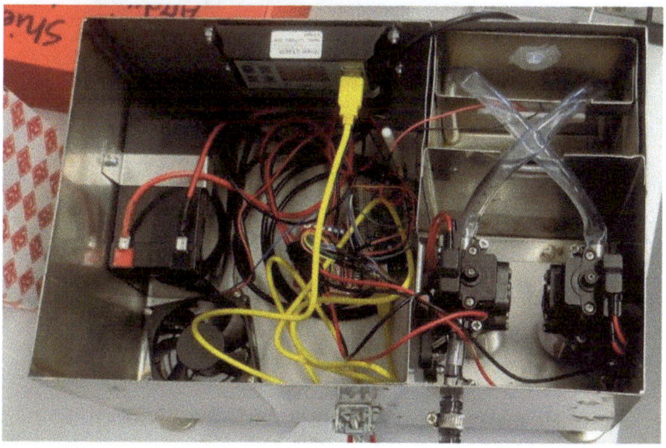

Fig. 7.7 Inside of the sensor node

LG01S [12] is used as the LoRa/Wi-Fi gateway to communicate between the sensing node and the cloud server. The gateway is also responsible for sending the data to the cloud server. Thingspeak is an IoT-based cloud server, which is free to use and can easily store data. It has been used to store all the real-time data which were collected from the sampling field.

Both the WiFi and the LoRa communication protocols were used to observe the resilience of the sensor nodes. The Arduino Uno WiFi contains a WiFi module, which is accountable for transmitting and receiving data through the WiFi gateway. An LG01S is also used which is used as a WiFi gateway. All the microcontrollers, sensors, rechargeable battery, inlet and outlet pumps, solar water reservoir, and charge controller are constrained in a steel enclosure, which is robust and more comfortable to install. Figure 7.6 indicates the materials, which are used in the sensing node. Figure 7.7 shows the inside of the steel box.

- Energy Harvesting Technique
  A solar panel (Model: ZM-9051), a sealed rechargeable battery (12 V, 12AH), and a solar charge controller (MP-3750) were used to deliver continuous energy without any human involvement. The proposed system is controlled from the microcon-

Fig. 7.8 Field installation of the smart sensor node

troller in numerous operational modes (active, sleep, transmitting/receiving) in all through the day. The steel enclosure was designed for housing all the electronics and the reservoirs. The steel structure was robust and more accessible to install in any sampling locations.
A VERT900 omnidirectional antenna was used as an external antenna and extended through the steel enclosure. Finally, it was installed to close to the study location as per Fig. 7.8.

- Study-Location
A small creek is available inside the Macquarie University campus and located as a study location. As it is seen from Fig. 7.9, the sensor node is situated 340 m away from the gateway. There should be a clear line-of-sight between the sensor node and the gateway to avoid any obstacle during the data transmission. The blue marker in the map designates the gateway location, and the green marker designates the sensor node's location.

7.5 Results and Discussions

- EIS Measurement of the Concentrations of Nitrate
The real and imaginary impedances for concentrations of nitrate samples are measured during an EIS measurement and plotted as Fig. 7.10. It is seen from Fig. 7.10 that the change of impedance of various concentrations of 1, 10, 20, 30, and 40 ppm are different from each other due to the different dielectric properties of the samples. When the concentrations of nitrate are different, the number of ions for also different and eventually their dielectric properties are also different. Therefore, when the sensing area dipped inside the water samples, the electric field from the positive electrode bulges through the ions in the aqueous medium towards

Fig. 7.9 Study location and distance between the gateway and the sensor node

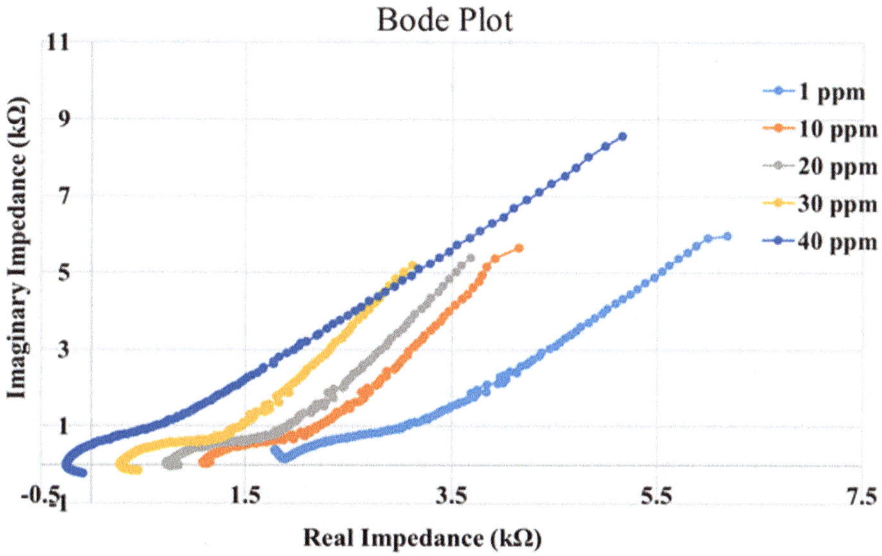

Fig. 7.10 Bode plots for different nitrate concentrations

the negative electrode. The resultant impedance behaviours are different, which is reflected in the bode plot [13–15].

The real impedances and imaginary impedances are plotted concerning frequency, and it is to be noted that the real impedance has a substantial change, contrasting the imaginary impedance. The change is unchanging and steady when the frequency is more than 500 Hz.

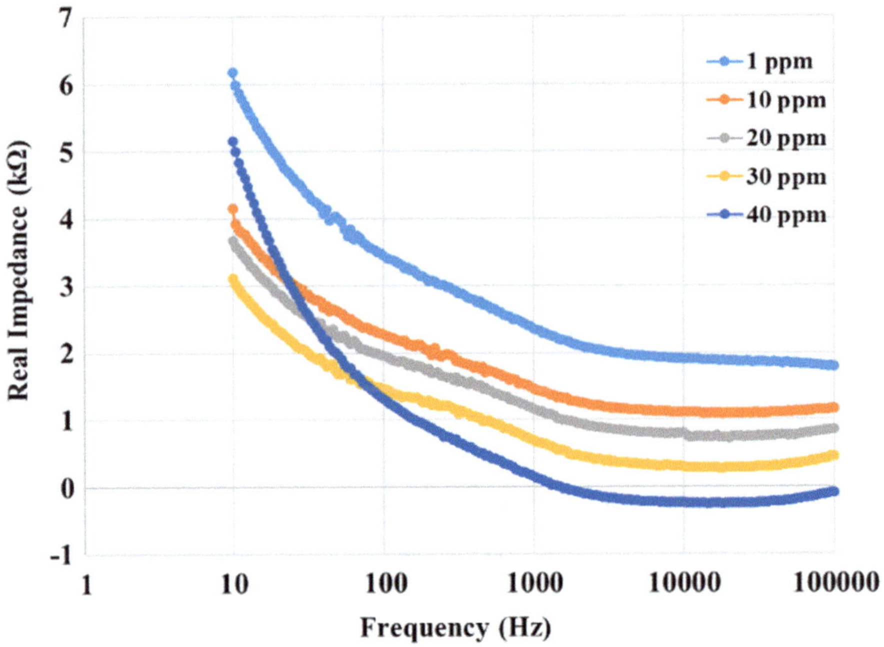

Fig. 7.11 Frequency versus real impedance for different nitrate concentrations

- Calibration Standard

  It is seen from Fig. 7.11 that the sensitive region for various concentrations is from 500 Hz to 100 kHz. There is a noteworthy change in the real impedance in that frequency range for different nitrate concentrations. 1000 Hz is preferred as the operating frequency to progress a calibration standard to measure any unidentified water samples which contain nitrate. At 1000 Hz, the real impedance of 1, 10, 20, 30, and 40 ppm of nitrate concentrations was used to grow a calibration standard to measure an unknown concentration. It is seen from Fig. 7.12 that the various nitrate concentrations are plotted on the x-axis and the analogous real impedances are plotted on the y-axis. They follow a linear straight line, and the regression coefficient is $\mathbf{R^2 > 0.98}$, which is satisfactory to calculate an unknown impedance and unknown nitrate concentrations from any water samples. It also means that the actual impedance and predicted impedance are close to each other.

  Therefore, the calibration standard for an unknown sample concentration is:

$$\boldsymbol{C} = \frac{\boldsymbol{R} - \mathbf{2212.3}}{\mathbf{-51.77}} \tag{7.2}$$

  where C is the concentration (ppm) and R (Ω) is the measured real impedance from an unknown sample. It is calculated that the sensitivity of the sensor is **−51.77 Ω/ppm**.

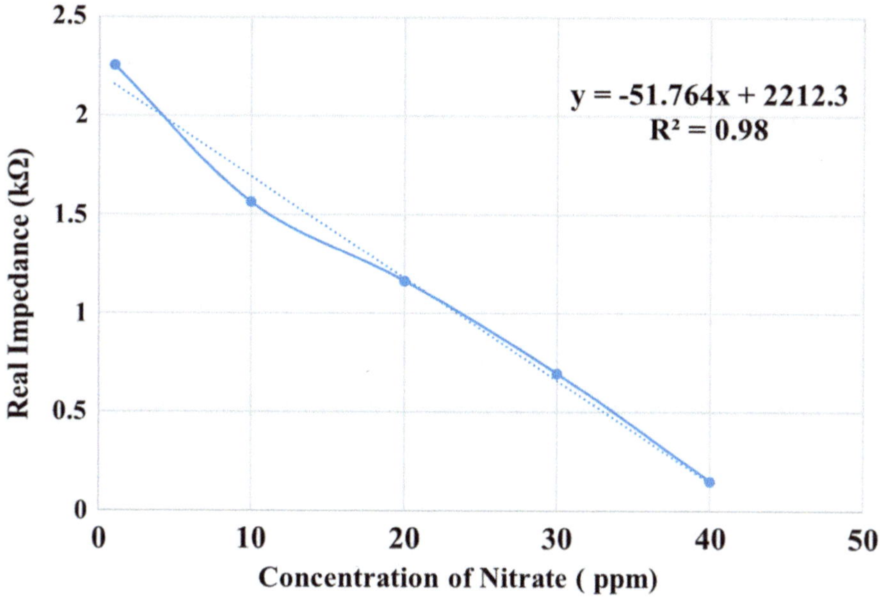

Fig. 7.12 Calibration standard to measure any unknown nitrate concentration

- Unknown Sample Measurement
  Various sample waters are collected from various sampling locations, such as river, lake, stream, tap water to consider measuring unknown samples. Unsurprisingly, the nitrate concentration was not high enough to measure the range to high concentrations. Therefore, a certain amount of nitrate sample was added to uplift the nitrate concentration. The proposed FR4-based sensor and smart sensing system were used to measure the concentration of nitrate and compared with the standard laboratory method. Equation 7.2 was used to measure the real impedance which is used to calculate the unknown nitrate concentrations. It is observed from Table 7.1 that the sensor and the proposed system can measure the concentration of nitrate with an error of less than 5%. When the nitrate concentrations are on the upper limit, the error is smaller than with lesser nitrate concentrations.
- Data Transfer to Cloud Server
  It is seen from Fig. 7.13 that the measured concentration of nitrate was transferred from the sampling field location during the on-field trial. Seven days of sampling data were collected without any disruption and the sampling time was 13 min. The sampling time is easily controllable through software programming, though this was a very dense data collection. 112–113 batches of on field sampling data could be collected with sampling every 13 min in 24 h, Table 7.2 shows the average concentration of nitrate and compares the data with standard laboratory measurements. The sample water was collected from the creek in every morning, afternoon and evening, and measured in the laboratory immediately. In a single

Table 7.1 Unknown sample measurement compared to laboratory standard method

Sample number	Smart sensing node (ppm)	UV-spectrometry method (ppm)	Error (%)
1. River water[a]	15.6	15.8	1.27
2. Tap water[a]	26.4	26.97	2.11
3. Canal water[a]	35.45	36.01	1.56
4. Stream water[a]	8.65	8.7	0.57
5. Creek water	2.01	2.1	4.29

[a]The nitrate concentrations are elevated for these samples

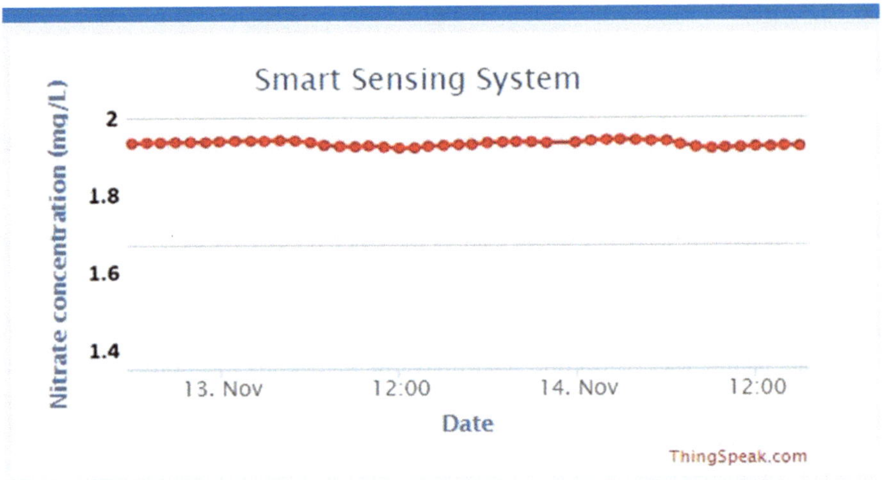

Fig. 7.13 Nitrate concentration is Thingspeak server

day, the sampling water was collected and measured in the laboratory to maximise accuracy. It was evident that the sampling frequency of collecting the sample water manually will not be analogous to the developed smart system. It is also observed that the measurement error (less than 5%) of the sampling data from the developed system is close to the laboratory measurement. A similar graph is also seen in Fig. 7.14.

Figure 7.15 illustrates the single-day nitrate concentrations for each sample time. It is varied between 1.8 and 1.9 ppm while the average concentration of nitrate of that single day was 1.9 ppm. Besides, a simple moving-average algorithm was used in the collected data to smooth the trend. The smoothed data can offer the trend of concentration of nitrate, which might be useful over a long time to monitor the nitrate concentration in real-time. The creek water does not carry an upper level of nitrate concentration as it is situated inside the Macquarie University

Table 7.2 Average concentration of nitrate of the study location compared with the laboratory standard method

Date	Smart sensing node (ppm)	UV-spectroscopy method (ppm)	Error (%)
8th Nov 2017	1.72	1.75	1.71
9th Nov 2017	2.01	1.95	3.07
10th Nov 2017	1.86	1.88	1.06
11th Nov 2017	1.98	1.96	1.02
12th Nov 2017	1.95	1.97	1.01
13th Nov 2017	1.9	1.92	2.60
14th Nov 2017	1.91	1.89	1.05

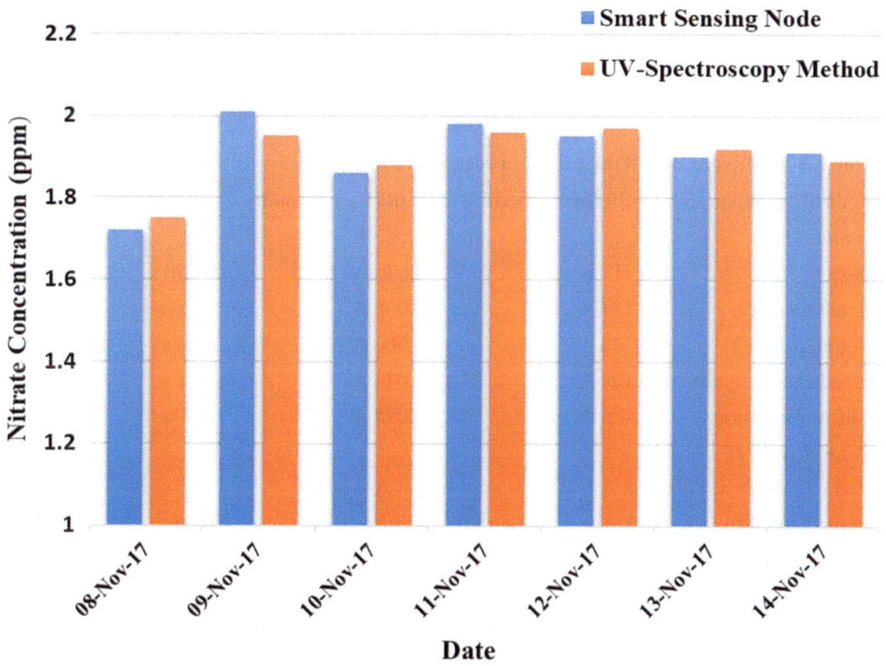

Fig. 7.14 Comparison of daily evolution of nitrate concentration with the standard method

and the management of the University regulates all kind of pollution (air, water, environment) very carefully.

- Comparison of Wi-Fi Protocol over LoRa protocol
  Electronic components for remote IoT applications must consume as little power as possible to maximise battery life. The difference of WiFi RF protocol over the LoRa protocol was to decrease total power consumption throughout sleep mode to less than half. It also permits data transmissions to penetrate various obstacles

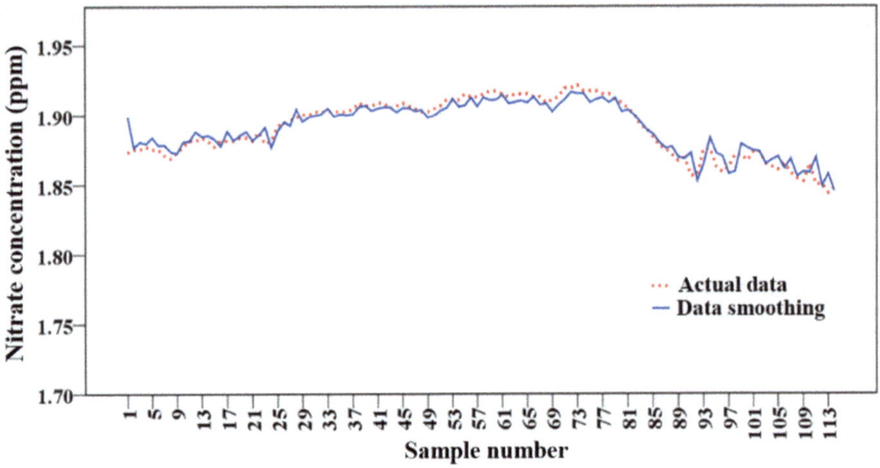

Fig. 7.15 Nitrate concentration of a single day

Table 7.3 Comparison of WiFi and LoRa based sensing system

Mode of operation	WiFi (per second)	LoRa (per second)	Per sample (13 min)
Arduino (5 V)			
Sleep	0.119 A	0.039 A	550 s
Motor 1 (inlet)	0.124 A	0.066 A	35 s
Motor 2 (outlet)	0.124 A	0.066 A	66 s
Impedance analyzer	0.124 A	0.046 A	18 s
Data transmission	0.124 A	0.095 A	10 s
Motor driver (12 V)			
Motor 1 (inlet)	1.8 A	1.8 A	35 s
Motor 2 (outlet)	1.8 A	1.8 A	66 s

and allows the collected data to travel more distances that are substantial whereas consuming a smaller amount of power than the standard WiFi protocol. The LoRa protocol optimises the exchange of data with the gateway, permitting for lower power consumption as compared to WiFi. The current consumption from each electronic component for each stage, linking WiFi and LoRa, can be detected in Table 7.3. It is observed that the current consumption for the LoRa protocol is less compare to WiFi, which helped to harvest the solar energy for an extended time. The energy consumption for inlet and outlet pumps and motor drivers are alike for both the systems. However, WiFi enabled systems to consume 2.22 times more energy than the LoRa enabled the system. Therefore, a LoRa enabled sensing system was proposed and trialled during the necessary field trial.

7.6 Chapter Summary

An IoT-enabled smart nitrate sensor and sensing system are proposed to monitor the concentration of nitrate in real-time. An FR4 interdigital capacitive sensor is characterised, and a calibration curve is developed to measure unknown sample from various sources of water. The system is self-governing and trialled in the field for seven days without any disruption. The LoRa protocol was used to run the system over more extended periods than for the WiFi protocol due to the power consumption variation. The LoRa protocol is a low-power energy-saving protocol, which is executed successfully. The proposed system's collected data were also validated through the standard UV-spectrometry method. The collected data displays that few data have lost throughout the transmission and 98% data are effectively collected through the gateway to the IoT based cloud server. The results show that the proposed smart sensing system can be beneficial to improve a WSN network to monitor the concentration of nitrate in real time. It also demonstrates that, without human help, changes of both time-based and spatial evolutions of the concentration of nitrate can be monitored successfully.

References

1. A.V. Mamishev, K. Sundara-Rajan, F. Yang, Y. Du, M. Zahn, Interdigital sensors and transducers. Proc. IEEE **92**(5), 808–845 (2004)
2. Y. Chen, C. Zhu, M. Cao, T. Wang, Photoresponse of SnO_2 nanobelts grown in situ on interdigital electrodes. Nanotechnology **18**(28), 285502 (2007)
3. S.C. Mukhopadhyay, C.P. Gooneratne, G.S. Gupta, S.N. Demidenko, A low-cost sensing system for quality monitoring of dairy products. IEEE Trans. Instrum. Meas. **55**(4), 1331–1338 (2006)
4. M.E.E. Alahi, N. Afsarimanesh, S. Mukhopadhyay, L. Burkitt, P.-L. Yu, In Highly selective ion imprinted polymer based interdigital sensor for nitrite detection, in *2016 10th International Conference on Sensing Technology (ICST)* (IEEE 2016), pp. 1–5
5. M.E.E. Alahi, N. Pereira-Ishak, S.C. Mukhopadhyay, L. Burkitt, An internet-of-things enabled smart sensing system for nitrate monitoring. IEEE Internet Things J. (2018)
6. J. de Carvalho Silva, J.J. Rodrigues, A.M. Alberti, P. Solic, A.L. Aquino, LoRaWAN—a low power WAN protocol for Internet of things: a review and opportunities, in *2017 2nd International Multidisciplinary Conference on Computer and Energy Science (SpliTech)* (IEEE 2017), pp. 1–6
7. M.T. Penella-López, M. Gasulla-Forner, *Powering autonomous sensors: an integral approach with focus on solar and RF energy harvesting* (Springer Science & Business Media, 2011)
8. Arduino Arduino Uno WiFi. https://store.arduino.cc/usa/arduino-uno-wifi. Accessed 09 April 2017
9. A. Devices, AD5933: Impedance Analyzer. http://www.analog.com/media/en/technical-documentation/data-sheets/AD5933.pdf. Accessed 09 April 2017
10. H. Electronic, RFM95/96/97/98(W)—low power long range transceiver module V1.0. http://www.hoperf.com/upload/rf/RFM95_96_97_98W.pdf. Accessed 13 Nov 2017
11. Sparkfun Dual Full Bridge Driver. https://www.sparkfun.com/datasheets/Robotics/L298_H_Bridge.pdf. Accessed 13 Nov 2017
12. Dragino LG01-S LoRa Gateway. http://www.dragino.com/products/lora/item/119-lg01-s.html. Accessed 13 Nov 2017

13. E. Ivers-Tiffée, A. Weber, H. Schichlein, Electrochemical impedance spectroscopy. *Handbook of fuel cells* (2010)
14. B.-Y. Chang, S.-M. Park, Electrochemical impedance spectroscopy. Ann. Rev. Anal. Chem. **3**, 207–229 (2010)
15. M.E. Orazem, B. Tribollet, *Electrochemical impedance spectroscopy,* vol 48 (Wiley, 2011)

Chapter 8
Conclusion and Future Work

8.1 Conclusion

Regular measurement of nitrate is critical to keep the water safe for all purposes. In this book, it is discussed about IoT enabled smart sensors and sensing system which were developed to measure nitrate concentration in water. The sensing system can be installed any sampling locations, and the system can measure the nitrate concentration and transfer it to the cloud server for further analysis. The purpose of this book can be concluded in the following manner.

Chapter 4 explains a temperature compensated interdigital capacitive sensor for measuring nitrate at low concentrations. A portable, novel sensing system has been developed that could be used on-site as a stand-alone device, as well as IoT-based remote monitoring smart sensor node, to measure nitrate concentration in surface and groundwater. Electrochemical Impedance Spectroscopy was employed to detect and display nitrate concentrations, by evaluating the impedance change read by the interdigital transducer immersed in the surface water samples. The test samples were evaluated by commercial equipment (LCR meter) and the designed system. These results were also validated using standard laboratory techniques to assess nitrate concentrations in water samples. The developed system has the potential to be used to estimate nitrate concentrations in water samples, in real-time. The system can upload the measured nitrate data on a website based on IoT. This system could be used to integrate water quality monitoring sites within farms, or between streams, rivers, and lakes.

Chapter 5 explains and highlights the fabrication process of a newly developed CNT/PDMS and Graphene sensor. It also explains the IoT-enabled sensing system for the detection of nitrate-N concentration in water. The graphene-based sensor is easy to fabricate and easier to use for measurement purpose. The sensor is also robust which helps to duplicate measurements with good repeatability. The developed sensor and the smart sensing system can be used to monitor real-time nitrate-N concentration and develop a low-cost distributed network with adequate performance.

Md E. E. Alahi and S. C. Mukhopadhyay, *Smart Nitrate Sensor*, Smart Sensors, Measurement and Instrumentation 35, https://doi.org/10.1007/978-3-030-20095-4_8

Chapter 6 explains the ion imprinted polymer-based nitrate-N has been developed and used as a coating material the sensing surface. The interdigital capacitive sensor has been used for detection purposes where the detection limit is 1–10 mg/L. The sensing surface was functionalized by an acrylic resin with an embedded coating material to introduce selectivity for nitrate-N. A calibration curve has been developed to measure an unknown sample. The measurement of unknown water samples has shown good results, with an error rate of <5%. The sensor can be reused up to 5 times before the error rate becomes unacceptable. CNLS curve fitting was also applied to interpret the experimentally obtained impedance spectra.

Chapter 7 explains an IoT-enabled smart nitrate sensor and sensing system to monitor the nitrate concentration in real-time. An FR4-based interdigital sensor is characterised and is extremely useful for robust use during nitrate monitoring. The system is autonomous and was trialled in the field for seven days without any interruption. The LoRa protocol was used to run the system over more extended periods than for theWiFi protocol. The LoRa protocol is a low-power energy-saving protocol, which was implemented successfully. The system's collected data were also validated through the UV-spectrometry method. The collected data shows that few data have been lost during transmission, and 98% of data are successfully collected through the gateway to the cloud server. The results show that the proposed smart sensing system can be beneficial to develop a WSN to monitor nitrate concentration in real time. It also shows that, without human interaction, changes of both temporal and spatial evolutions of nitrate concentration can be monitored successfully.

8.2 Future Work

Satisfactory performance of the sensor and sensing system has been explained in this book. Still, a few of the following improvements in the sensor and sensing system can make it even better.

1. The reuse of the sensor is an essential parameter which needs to be improved.
2. There are other heavy metals, and concerned ions are available in water. Sensor arrays can be developed to monitor other water parameters.
3. The material of the sensors is significant. The sensor's life can be dependent on the material of the sensor. Therefore, more research should be done on sensor materials to improve the life of the sensors.
4. The size and cost of the sensing system can be reduced.
5. Data Security is one of the critical parameters for IoT enabled sensing system. The security is not discussed in this book. Therefore, data security can be introduced into the developed sensing system.

MIX
Papier aus verantwortungsvollen Quellen
Paper from responsible sources
FSC® C105338
FSC
www.fsc.org

If you have any concerns about our products,
you can contact us on
ProductSafety@springernature.com

In case Publisher is established outside the EU,
the EU authorized representative is:
Springer Nature Customer Service Center GmbH
Europaplatz 3, 69115 Heidelberg, Germany

Printed by Libri Plureos GmbH
in Hamburg, Germany